aus der Reihe:

Innovationen mit Mikrowellen und Licht

Forschungsberichte aus dem Ferdinand-Braun-Institut, Leibniz-Institut für Höchstfrequenztechnik

Band 60

Tasmim Alam

Spectroscopic Applications of Terahertz Quantum-Cascade Lasers

Herausgeber: Prof. Dr. Günther Tränkle, Prof. Dr.-Ing. Wolfgang Heinrich

Ferdinand-Braun-Institut
Leibniz-Institut
für Höchstfrequenztechnik (FBH)
Gustav-Kirchhoff-Straße 4
12489 Berlin

Tel. +49.30.6392-2600
Fax +49.30.6392-2602

E-Mail fbh@fbh-berlin.de
Web www.fbh-berlin.de

Innovations with Microwaves and Light

Research Reports from the Ferdinand-Braun-Institut, Leibniz-Institut für Höchstfrequenztechnik

Preface of the Editors

Research-based ideas, developments, and concepts are the basis of scientific progress and competitiveness, expanding human knowledge and being expressed technologically as inventions. The resulting innovative products and services eventually find their way into public life.

Accordingly, the *"Research Reports from the Ferdinand-Braun-Institut, Leibniz-Institut für Höchstfrequenztechnik"* series compile the institute's latest research and developments. We would like to make our results broadly accessible and to stimulate further discussions, not least to enable as many of our developments as possible to enhance everyday life.

This work describes the development of novel methods for high-resolution terahertz (THz) molecular spectroscopy with quantum-cascade lasers (QCLs). It is the first time that the observation of a Lamb-dip is reported when using a THz QCL. Another obstacle which has been successfully addressed in this work is the frequency tunability of QCLs. By near-infrared illumination of one facet of the QCL a record frequency tunability of 40 GHz has been obtained. The results of this work pave the way for high-resolution molecular spectroscopy beyond the Doppler limit with THz QCLs.

We wish you an informative and inspiring reading

Prof. Dr. Günther Tränkle
Director Ferdinand-Braun-Institut,
Leibniz-Institut für Höchstfrequenztechnik

Prof. Dr. Heinz-Wilhelm Hübers (guest editor)
Director Institute of Optical Sensor Systems
German Aerospace Center (DLR)

The Ferdinand-Braun-Institut

The Ferdinand-Braun-Institut researches electronic and optical components, modules and systems based on compound semiconductors. These devices are key enablers that address the needs of today's society in fields like communications, energy, health and mobility. Specifically, FBH develops light sources from the visible to the ultra-violet spectral range: high-power diode lasers with excellent beam quality, UV light sources and hybrid laser systems. Applications range from medical technology, high-precision metrology and sensors to optical communications in space and integrated quantum technology. In the field of microwaves, FBH develops high-efficiency multi-functional power amplifiers and millimeter wave frontends targeting energy-efficient mobile communications as well as car safety systems. In addition, compact atmospheric microwave plasma sources that operate with economic low-voltage drivers are fabricated for use in a variety of applications, such as the treatment of skin diseases.

The FBH is a competence center for III-V compound semiconductors and has a strong international reputation. FBH competence covers the full range of capabilities, from design to fabrication to device characterization.

In close cooperation with industry, its research results lead to cutting-edge products. The institute also successfully turns innovative product ideas into spin-off companies. Thus, working in strategic partnerships with industry, FBH assures Germany's technological excellence in microwave and optoelectronic research.

Spectroscopic Applications of Terahertz Quantum-Cascade Lasers

vorgelegt von

M. Sc.

Tasmim Alam

geboren in Rajshahi, Bangladesch

von der Fakultät IV – Elektrotechnik und Informatik
der Technischen Universität Berlin zur Erlangung
des akademischen Grades

Doktor der Ingenieurwissenschaften

Dr.-Ing

genehmigte Dissertation

Vorsitzender: Prof. Dr.-Ing. Lars Zimmermann

Gutachter: Prof. Dr. Günther Tränkle

Gutachter: Prof. Dr. Heinz-Wilhelm Hübers

Gutachter: Prof. Dr. Martin R. Hofmann

Tag der wissenschaftlichen Aussprache: 17 August, 2020

Berlin, 2020

Abstract

Quantum cascade lasers (QCLs) are attractive for high-resolution spectroscopy because they can provide high power and a narrow linewidth. They are particularly promising in the terahertz (THz) range since they can be used as local oscillators for heterodyne detection as well as transmitters for direct detection. However, THz QCL-based technologies are still under development and are limited by the lack of frequency tunability as well as the frequency and output power stability for free-running operation. In this dissertation, frequency tuning and linewidth of THz QCLs are studied in detail by using rotational spectroscopic features of molecular species. In molecular spectroscopy, the Doppler effect broadens the spectral lines of molecules in the gas phase at thermal equilibrium. Saturated absorption spectroscopy has been performed that allows for sub-Doppler resolution of the spectral features. One possible application is QCL frequency stabilization based on the Lamb dip. Since the tunability of the emission frequency is an essential requirement to use THz QCL for high-resolution spectroscopy, a new method has been developed that relies on near-infrared (NIR) optical excitation of the QCL rear-facet. A wide tuning range has been achieved by using this approach. The scheme is straightforward to implement and, the approach can be readily applied to a large class of THz QCLs. The frequency and output stability of the local oscillator has a direct impact on the performance and consistency of the heterodyne spectroscopy. A technique has been developed for a simultaneous stabilization of the frequency and output power by taking advantage of the frequency and power regulation by NIR excitation. The results presented in this thesis will enable the routine use of THz QCLs for spectroscopic applications in the near future.

Zusammenfassung

Quantenkaskadenlaser (QCLs) sind für die hochauflösende Spektroskopie attraktiv, da sie eine hohe Leistung und eine schmale Linienbreite bieten können. Sie sind im THz-Bereich besonders vielversprechend, da sie als lokale Oszillatoren (LO) für Heterodynspektroskopie sowie als Strahlungsquellen für direkte Detektion eingesetzt werden können. THz-QCL-basierte Technologien befinden sich jedoch noch in der Entwicklung und sind durch die mangelnde Frequenzabstimmbarkeit sowie die Frequenz- und Ausgangsleistungsstabilität im frei laufenden Betrieb begrenzt. In dieser Dissertation werden Frequenzabstimmung und Linienbreite von THz-QCLs unter Verwendung rotationsspektroskopischer Merkmale molekularer Spezies detailliert untersucht. Im thermischen Gleichgewicht verbreitert der Doppler-Effekt die Spektrallinien von Molekülen in der Gasphase. Mit Hilfe von Sättigungsspektroskopie konnte in dieser Arbeit eine Auflösung der Spektralmerkmale unterhalb des Doppler-Limits erreicht werden. Eine mögliche Anwendung ist die Lamb-Dip-basierte Frequenzstabilisierung. Eine wesentliche Voraussetzung für die Verwendung von THz QCL für hochauflösenden Spektroskopie ist die Abstimmbarkeit der Emissionsfrequenz. Hierfür wurde eine neue Methode entwickelt, die auf der optischen Anregung der QCL-Rückseite im nahen Infrarot (NIR) beruht. Mit diesem Ansatz wurde ein breiter Abstimmbereich erreicht. Das Schema ist einfach zu implementieren und kann leicht auf eine große Klasse von THz-QCLs angewendet werden. Da die Frequenz- und Ausgangsstabilität des lokalen Oszillators sich unmittelbar auf die Auflösung und Empfindlichkeit eines Heterodynspektrometers auswirkt, wurde eine Technik zur gleichzeitigen Stabilisierung von Frequenz und Ausgangsleistung entwickelt. Hierfür wurde die Frequenz- und Leistungsregelung durch NIR-Anregung ausgenutzt. Die in dieser Arbeit vorgestellten Ergebnisse werden in naher Zukunft die häufige Verwendung von THz-QCLs für spektroskopische Anwendungen ermöglichen.

List of Abbreviations

AC	Alternating current
BWO	Backward wave oscillator
cw	Continuous-wave
DC	Direct current
DL	Diode laser
DAQ	Data Acquisition
DFG	Difference frequency generation
FWHM	Full width at half maximum
FMS	Frequency modulation spectroscopy
FTS	Fourier transform spectroscopy
FEL	Free-electron laser
GREAT	German Receiver for Astronomy at THz Frequencies
GaAs	Gallium arsenide
HEB	Hot electron bolometer
HDPE	High-density polyethylene
IR	Infrared
LAS	Laser absorption spectroscopy
LIFT	Light-induced frequency tuning
L-I-V	Light-current-voltage
LO	Local oscillator
MBE	Molecular beam epitaxy
MOCVD	Metal organic chemical vapor deposition

MM	Metal-metal
NIR	Near-infrared
ND	Natural density
NEP	Noise equivalent power
OD	Optical density
PCA	Photoconductive antenna
PID	Proportional integral derivative
PSD	Power spectral density
QCL	Quantum-cascade laser
RMS	Root mean square
RT	Room temperature
SOFIA	Stratospheric Observatory for Infrared Astronomy
SP	Single plasmon
SIS	Superconductor-insulator-superconductor
THz	Terahertz
TPX	Transparent polymer X (poly-4-methylepentene-1)
TUNNET	Tunnel injection transit-time
WMS	Wavelength modulation spectroscopy

Symbols

α	absorption coefficient
α_0	unsaturated absorption coefficient
β	phase modulation
η_i	injection effciency
ε_0	vacuum permitivity
λ	wavelength
ν	frequency
ν_{LO}	local oscillator frequency
ν_m	modulation frequency
ν_0	center frequency
ν_s	signal frequency
$\Delta\nu_d$	Doppler broadening
$\Delta\nu_p$	pressure broadening
γ_L	damping factor
ω	angular frequency
ω_p	plasma frequency
ϕ	phase
σ	absorption cross section
τ_{nr}	non-radiative lifetime
τ	lifetime
Γ	confinement factor
θ	angle
a_w	waveguide loss
A	area
B_r	bimolecular recombination
c	speed of light
e	charge of electron
E	electric field
g	spacing

G	generation rate
h	Planck's constant
I_0	unattenuated pump intenisty
I_{pump}	pump intenisty
I_{probe}	probe intenisty
I_s	saturation intenisty
J	current density
k	Boltzmann constant
k_l	wave vector
L	length
m^*	effective electron mass
M	molecular weight
M^2	M square value
N	density of molecules
p	pressure
P	power
S	saturation parameter
S_0	normalized incident photon flux
t	time
T	temperature
V	velocity

Contents

Abstract i

Zusammenfassung iii

List of Abbreviations v

Symbols vii

1 Introduction 1

1.1 Terahertz radiation 1
1.2 Applications of THz radiation 3
1.2.1 Security screening 3
1.2.2 Biomedical research 3
1.2.3 Astrophysics and atmospheric sensing 4
1.2.4 Wireless communication 4
1.3 Organization of thesis 5

2 THz sources, detectors, and components 7

2.1 THz sources 7
2.1.1 Solid-state sources 7
2.1.2 Electron beam sources 8
2.1.3 THz gas laser 8
2.1.4 Difference frequency generation 9
2.2 THz detectors 10
2.2.1 Ge:Ga photoconductive detectors 10
2.2.2 Golay cell detectors 11
2.2.3 Microbolometers 12
2.2.4 Pyroelectric detectors 13
2.3 THz optical components 14
2.3.1 THz windows 14

2.3.2 THz optics . . . 15
2.3.3 Polarizers . . . 16
2.3.4 Waveplates . . . 16

3 Fundamentals of terahertz quantum-cascade lasers and laser absorption spectroscopy 19
3.1 Historical overview of QCL . . . 19
3.2 Active region of QCL . . . 20
3.3 Active region designs . . . 22
3.4 Waveguides . . . 23
3.5 Temperature performance of THz QCLs . . . 25
3.6 Spectroscopic techniques . . . 25
3.6.1 Laser absorption spectroscopy . . . 25
3.6.2 Heterodyne spectroscopy . . . 30
3.6.3 Modulation spectroscopy . . . 31
3.6.4 Saturation spectroscopy . . . 32

4 Experimental techniques 35
4.1 Cooling system . . . 35
4.1.1 Helium flow cryostat . . . 35
4.1.2 Mechanical cryocooler . . . 36
4.2 Power measurement of THz QCLs . . . 37
4.3 Light-current-voltage characterization . . . 38
4.4 Beam characterization . . . 39
4.5 Fourier transform spectroscopy . . . 41
4.6 High-resolution molecular spectroscopy with a THz QCL . . . 42
4.6.1 Measurement setup . . . 43
4.6.2 Data Acquisition . . . 44
4.6.3 Frequency calibration . . . 45

5 Doppler free spectroscopy with a THz QCL 47
5.1 Theoretical background . . . 48
5.2 Experimental setup for QCL characterization and saturated absorption spectroscopy . . . 49
5.3 Results . . . 50
5.4 Experimental setup for Lamb-dip spectroscopy . . . 52
5.5 Results . . . 53
5.6 Doppler-free spectroscopy inside a mechanical cryocooler . . . 55

5.7 Conclusion . . . 57

6 Molecular spectroscopy by light-induced frequency tuning of THz QCLs 59

6.1 Theoretical background . . . 60

6.2 Experimental setup . . . 66

6.2.1 Setup for frequency tuning . . . 67

6.2.2 Setup for molecular spectroscopy . . . 68

6.3 Results . . . 68

6.4 Illumination in different regions of the QCL . . . 70

6.5 Comparison of light-induced, current and temperature tuning effects . . . 71

6.6 Frequency tuning by continuously variable natural density filter . . . 75

6.7 Future prospects . . . 76

7 Frequency and output power stabilization of a THz QCL using near-infrared illumination 79

7.1 Experimental setup . . . 80

7.2 Frequency tuning measurements . . . 81

7.3 Frequency modulation spectroscopy . . . 82

7.4 Frequency and output power stabilization . . . 83

7.5 Frequency stabilization with NIR laser and power stabilization with QCL driving current . . . 88

7.6 Longterm measurement . . . 90

7.7 Conclusion . . . 91

8 Summary and outlook 93

Bibliography 95

Acknowledgements 111

List of Figures

1.1 Electromagnetic spectrum and THz applications 2
1.2 Security screening applications: THz images of various hidden objects . . 3

2.1 Lasing process of an optically excited THz gas laser 9
2.2 Ge:Ga Photoconductive detectors 11
2.3 Design of a Golay Cell Detector 12
2.4 Microbolometer structure 13
2.5 Schematic diagram of a pyroelectric detector element 13
2.6 THz collimating optics 15
2.7 Schematic of a WG polarizer 16

3.1 Schematic illustration of lasing and cascading in the superlattice structure of the QCL 20
3.2 Schematic diagram of the active region of THz QCLs 22
3.3 Design and mode profile for SP waveguides and MM waveguides for THz QCL 24
3.4 Simplified setup of laser absorption spectroscopy. 26
3.5 Schematic setup for a heterodyne receiver. 30
3.6 Schematic setup for a frequency modulation spectroscopy. Left panel: direct signal, right panel: 1f signal. 32
3.7 Schematic setup for saturation spectroscopy. 33

4.1 THz QCL inside a He-flow cryostat. 36
4.2 THz QCL mounting inside a mechanical cryocooler. 37
4.3 Typical absolute power measurement setup for THz QCL. 38
4.4 Light-current-voltage characterization of THz QCL 39
4.5 Experimental setup for beam profile characterization. 40
4.6 Beam profile characterization of the THz QCL 41
4.7 Fourier transform spectroscopy based on THz QCL 42
4.8 Schematic of the experimental setup for QCL-based spectrometer. 43

4.9 Signal transmitted through the absorption cell filled with CH_3OH at a pressure of 1 hPa as a function of the driving current and heat-sink temperature of the QCL 44
4.10 Transmission signal due to current and temperature tuning 45
4.11 Frequency dependence due to current and temperature tuning 46

5.1 Schematics setup for linear absorption spectroscopy and saturation measurements 49
5.2 Spectral measurement of CH_3OH and compared with a JPL simulation, and absorption spectra for different isotopic compositions 50
5.3 Normalized absorption coefficient as a function of the normalized pump intensity for different values of the total pressure 51
5.4 Intensity profile of the QCL at the beam waist position. 52
5.5 The schematic diagram and the photograph of the lab setup used for Lamb dip spectroscopy 53
5.6 Normalized absorption signal for different intensities of the pump beam . 54
5.7 Lamb dip spectroscopy in the presence of optical feedback 56
5.8 Absorption line of HDO exhibiting a pronounced Lamb dip and absorption signal as a function of the driving current. 57

6.1 Calculated change of the real part of the refractive index close to the facet of the excited GaAs substrate for normalized incident photon fluxes . . . 61
6.2 Resonator with a length of L 62
6.3 Calculated shift of the resonance frequency as a function of intensity for a one-dimensional cavity 66
6.4 The experimental setup for LIFT technique 67
6.5 Detector signal, transmission spectrum after frequency calibration and frequency tuning characteristics for light-induced frequency tuning 69
6.6 Frequency tuning behavior due to optical excitation at the different positions of the substrate. (a) Schematic image of the facet of QCL Illuminated by NIR laser. (b) and (c) show the frequency shifts in the different substrate positions. 70
6.7 Frequency tuning behavior after NIR excitation in the substrate and the active region of the QCL. 71
6.8 Experimental results of QCL 2.5 THz for different modes 73
6.9 CH_3OH transmission spectrum of a 0.66 mm long 3.1 THz QCL, and normalized QCL intensity as a function of the normalized NIR power . . 74

6.10 Frequency tuning characterization results for continuously variable natural density filter 75

7.1 The schematic diagram and the photograph of the lab setup used for the frequency and output power stabilization 80
7.2 Frequency tuning measurements by varying the QCL driving current, and the diode laser intensity 81
7.3 Frequency modulation spectroscopy by modulating the driving current of the QCL or diode laser 83
7.4 Transmission signal and frequency-modulated signal of CH_3OH 83
7.5 Schematic block diagram of the PI control loop for the laser stabilization system. 84
7.6 Frequency stability versus time detected by the lock-in amplifier in the locking range indicated for three different cases are shown: (1) free running, (2) frequency stabilized, and (3) both, frequency and amplitude, stabilized 85
7.7 Dependence of the output frequency of the THz QCL by changing the set-point at the different point of the absorption line through the PID software system 86
7.8 Histograms of the frequency and power fluctuations corresponding to the unlocked state and the locked states 87
7.9 Power spectral density of the frequency noise and amplitude noise for the free-running, the frequency-locked, and the frequency and power-locked QCL 88
7.10 Frequency stability versus time detected by the lock-in amplifier for three different cases are shown: (1) free running, (2) frequency stabilized, and (3) both, frequency and amplitude, stabilized 89
7.11 Histograms of the frequency and power fluctuations corresponding to the unlocked state and the locked states 90
7.12 Longterm frequency and output power stabilization measurement 91

List of Tables

2.1 Optical constants and suitable frequency range for HDPE, TPX and silica quartz 14

6.1 Current and temperature tuning parameters for the investigated QCLs . . 72
6.2 Single-mode frequency coverage by LIFT for the three QCLs. 74

Chapter 1

Introduction

Any matter in this universe is made up of particles called atoms or molecules. Research on the interaction between light and matter using spectroscopic techniques provides great flexibility to probe the matter in new ways. When electromagnetic radiation interacts with matter, it involves either absorption, emission, or dispersion of radiation by the system being studied. Absorption or emission of atomic and molecular spectra can provide detailed information about the structure and chemical properties of the system. Spectroscopy has thus produced a significant contribution to the current state of atomic and molecular physics, to chemistry, and molecular biology. It is used in material research to identify different materials due to their intrinsic spectral fingerprints. In principle, one can identify spectral fingerprints of media that rely on the transitions from the ground state to the excited energy states of the medium. The accurate knowledge of the laboratory transition frequencies of the molecules using diagnostic tools is the prerequisite for the analysis of astronomical observation. A detailed analysis of the data yields valuable information on molecular parameters and is essential for the prediction of previously unmeasured transitions, either in frequency regimes not yet explored experimentally or of intensities not in the range of even the most sensitive spectroscopic techniques.

1.1 Terahertz radiation

The terahertz (THz) range, from 1 to 10 THz (30–300 μm) of the electromagnetic spectrum lies between microwave and infrared frequency range just at the border between photonics and electronics [c.f. Fig. 1.1]. This frequency spectrum of electromagnetic waves has broad potential in a wide range of applications, from fundamental research, such as molecular spectroscopy and astronomy, to practical areas such as environmental science, biomedical and security applications. A significant reason for the increased interest in THz research comes from astronomy. Many important astronomical molecules and atoms have their fingerprint absorption and emission spectra in the THz frequency range. To date, around 200 molecular species have been detected in the interstellar medium (ISM) and circumstellar shells [1]. Nevertheless, the generation and detection of THz waves are not as mature as in other regions of the electromagnetic spectrum. This frequency range has proven to be one of the most challenging to operate because the frequency is typically too high for traditional electronics, and the photon energy is too low compared to visible and near-infrared (NIR) light. Therefore, it is difficult to control and manipulate THz radiation using existing electronics and optics. As a result, the large portion of the THz range was not mainly explored because there were no suitable emitters to send controlled THz signals or active sensors to collect and record information. Over the last two decades, intensive research and development activities in academia and industry have tried to bridge the gap between microwave and infrared spectra. New concepts, techniques, and methods have been applied for the generation and detection of THz radiation in recent years. As a consequence, compact THz sources and detectors are now being developed that are capable of generating, detecting, and manipulating THz signals. Recent innovations allowing both robust and reliable THz sources, combined with high-performance THz spectroscopy and imaging systems, have opened up remarkable new opportunities in science and technology.

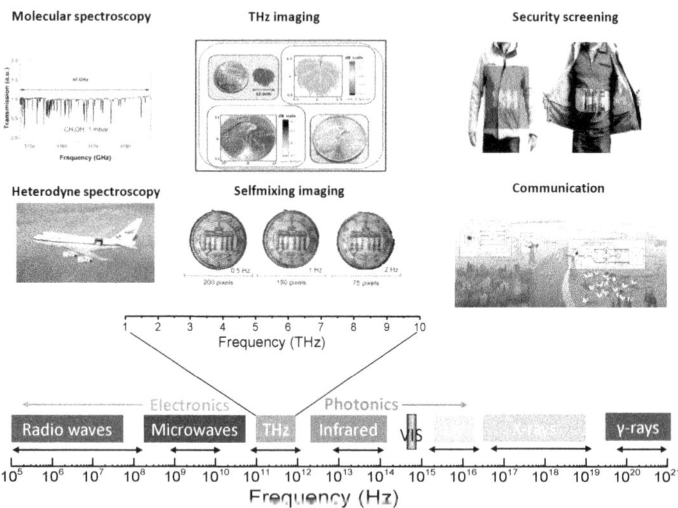

Figure 1.1: Electromagnetic spectrum and THz applications [2, 3, 4, 5, 6, 7].

1.2 Applications of THz radiation

The THz frequency spectrum is suitable for a variety of applications in different fields. Some of the very highlighted applications are discussed briefly in the following subsections.

1.2.1 Security screening

The development of techniques for the inspection of plastic explosives, chemical, and biological weapons has become essential as public safety concerns have increased considerably in recent years. THz imaging takes advantage of relatively low levels of non-ionizing THz radiation to detect hidden objects in fabrics, ceramics, and paper. It can reasonably handle security screening tasks such as checking mail packages, envelopes, and small parcels for hidden objects and threats. Figures 1.2 (a)–(e) display samples of THz images of various objects hidden under clothes, paper, plywood, and wood. Being organic compounds, most explosives have their unique spectral features in the THz band due to their rotation and collective vibrational transitions. Using THz spectroscopy, it should be possible to detect and identify unknown substances through their transmission characteristics and reflective spectra.

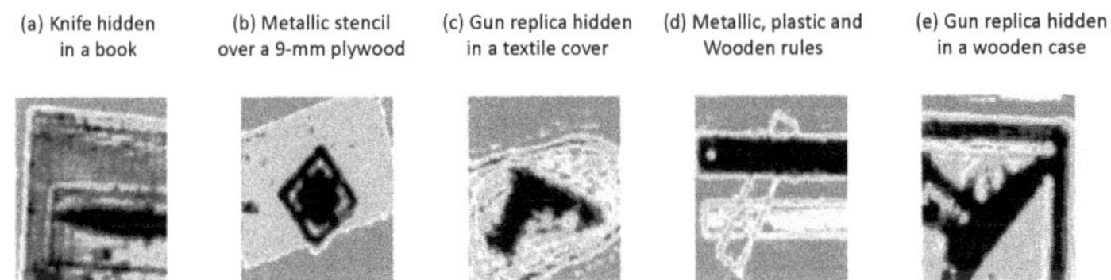

Figure 1.2: (a)–(e) Security screening applications: THz images of various hidden objects [6].

1.2.2 Biomedical research

THz waves have been used for the research of biological and medical applications. The investigation of biological features using THz radiation has both advantages and disadvantages. One of the advantages of THz radiation is the low photon energy, which is well below the ionization energies of atoms and molecules. This implies that one can study materials with THz radiation without any tissue damage unless too much power is applied. The other advantage is that biological molecules depend on hydrogen bonding, which is the most dominant bond in biological samples. For example, interactions between water and biomolecules cause hydrogen bonds to mediate and participate in a wide range of biomolecular interactions. THz radiation capable of detecting such spectral features of resonance and motion of molecules. The disadvantage of THz radiation is the high absorption in water in which biological molecules reside. Several biomedical studies have been performed in the THz frequency range. Among these studies are the determination of glucose concentration [8], monitoring DNA junction structures [9], breath gas analysis [10], water concentration determination [11], as well as many more.

1.2.3 Astrophysics and atmospheric sensing

Astronomy and astrophysics are the significant applications for continuous-wave (cw) THz spectroscopy. The THz spectral regime provides a wide range of spectral lines that are invaluable probes for star formation and planetary atmospheres [12]. It contains a plethora of spectroscopic features of atoms and molecules that are essential diagnoses of both the physical and chemical conditions of the gas and the energy sources in the astronomical environments. These spectral probes include rotational lines from simple molecules (e.g., HD, CH, OH, CO, H_2O, NH_3) and ground state fine structure lines from abundant atoms and ions (e.g. C, C^+, N^+, N^{++}, O, and O^{++}). As the wavelength of THz radiation is much larger than the typical size of the dust grains ($\sim$0.1 μm), the THz radiation is less scattered by interstellar dust as in the case for mid-infrared or optical radiation [12]. Observations in different ranges of the electromagnetic spectrum provide full information about the studied astronomical objects to the astronomers. One of the examples of astronomical observation is the Stratospheric Observatory for Infrared Astronomy (SOFIA), where the multi-frequency channels instrument GREAT (German Receiver of Astronomy at THz Frequencies) is used to acquire high-resolution spectra. The 4.7 THz heterodyne spectrometer system in the GREAT instrument has partly been developed in the German Aerospace Center, Berlin. This spectrometer system covers the frequency range around the fine structure line of the neutral atomic oxygen at 4.7448 THz [13].

1.2.4 Wireless communication

Over the past ten years, the rise of wireless devices and the rapid increase in data traffic have drastically increased the demand for spectral bandwidth, along with much faster data rates. Researchers, therefore, believe that switching to the carrier frequency above 0.1 THz could be a promising way to address such a huge demand. Numerous articles on various aspects of THz wireless communication measurements have been published in recent years. Carrier frequencies about 300 GHz are the most widely observed bands due to the availability of transmitter devices and components with sufficient output power.

To date, GaAs/InP electronics technologies are leading over 0.1 THz in all-electronic THz communication. They can cover frequency bands up to 300 GHz with a data rate of over 64 Gbit/s [14]. However, other semiconductor-based technologies have also been demonstrated for THz communication in the last few years. Si-electronics-based transceivers now easily enable 10 Gbit/s wireless links up to 260 GHz bands [15]. Communication links enabled by THz photonics technologies could play a significant role in the realization of an efficient THz wireless system. The highest data rate of about 100 Gbit/s for photonics-based transmitters has been demonstrated [16]. However, the transmission distance is limited by atmospheric attenuation as THz undergoes significant atmospheric absorption.

1.3 Organization of thesis

The main objective of this thesis is to explore spectroscopic applications of THz QCLs. The first introductory chapter provides an overview of the THz spectrum and its applications. In the second chapter, the fundamental overview of the sources available for THz radiation will be addressed. Besides, the detectors used in this dissertation and the components, including mirrors, windows, polarizers, and many others, will be discussed. The third chapter provides a basic overview of the working principle, the waveguide technique, and the temperature performance of the THz QCLs. The fundamentals of laser absorption spectroscopy, as well as its principle and nomenclature, will also be addressed here. In addition, various techniques for high-resolution spectroscopy will be briefly presented. The fourth chapter will then discuss the technical and experimental methods, including the cooling systems used for THz QCLs and the fundamental characterization techniques. Finally, the high-resolution molecular spectroscopy with THz QCLs will be discussed. Chapter 5 will present the technique and results for Doppler-free spectroscopy. The main objective of this type of spectroscopy is to study spectral features below the temperature-imposed Doppler linewidth limit. In Chapter 6, the wide-band frequency tuning of THz QCLs with a near-infrared (NIR) optical excitation will be addressed. The feasibility of this approach for molecular laser absorption spectroscopy is demonstrated in this work. Finally, Chapter 7 will demonstrate the technique for stabilizing the frequency and output power of a THz QCL. The technique exploits frequency and power variations upon near-infrared illumination of the QCL with a diode laser.

Chapter 2

THz sources, detectors, and components

This chapter provides a brief introduction to THz sources, detectors, and passive components. In section 2.1, the sources others than THz quantum cascade laser (QCL) are briefly presented. The detectors used for this thesis will be discussed in section 2.2. The materials used for passive THz components (lenses, mirrors, waveplates) will be addressed shortly in the last section.

2.1 THz sources

The THz generation process can be divided into two categories: the direct and the indirect generation process. Sources that directly generate THz waves through oscillation in electronic (electron beam or solid-state sources) or optical devices (QCLs) are known as direct processes. On the other hand, sources that generate THz radiation by photo-mixing in a non-linear medium or a medium with accelerated electrons are referred to as indirect processes. Sources that are relevant competitors of the THz QCL (frequency range, output power, and operating principle) will be briefly discussed in the following paragraphs.

2.1.1 Solid-state sources

Electronic solid-state sources such as oscillators and amplifiers are generally limited in frequency due to the transit time of carriers through semiconductor junctions, which cause the high-frequency roll-off. The output power drops off as $1/\nu^{\alpha}$ as the frequency ν increases. The values of α are between 1 and 3 [17]. Solid-state sources include resonant tunneling diodes (RTD), Gunn or transferred electron devices (TED), and transit time devices such as impact avalanche transit time (IMPATT) diodes and tunnel injection transit-time (TUNNETT) diodes. They are compact devices and can operate at room temperature. Gunn devices that are generating 0.2–50 μW power at 400–560 GHz frequency range are now feasible [18]. TUNNETT diodes with operational frequencies as high as 355 GHz with 140 μW output power have been reported [19]. In a THz frequency multiplier, the frequency of a driver source is multiplied in a nonlinear device to generate higher-order harmonic frequencies up to 2 THz. Planar Schottky diodes are commonly used in frequency multipliers, taking advantage of GaAs substrate-less technology to reduce the loss in the substrate.

2.1.2 Electron beam sources

Gyrotrons, free-electron lasers (FELs) and backward wave oscillators (BWOs) are electron beam sources capable of generating comparatively high output power, high-frequency THz radiation. All of these instruments are based on the interaction of a high-energy electron beam with a powerful magnetic field inside a resonant cavity or a waveguide, which results in an energy transfer between the electron beam and an electromagnetic wave. For gyrotrons, a power of 5 kW at 1 THz was demonstrated in pulsed mode [20] and 100 W were obtained for cw gyrotrons at frequencies ranging from 0.2 to 0.5 THz [21]. The FEL can provide broad frequency tuning, ultra-short pulses, and very high intensity in the THz region. BWOs can be operated in the THz region with moderate output power levels. They can be electrically tuned over a bandwidth of more than 50% of their operational frequencies and can generate a few mW at 1 THz [22]. The primary drawbacks of such a system are a large size, high cost, and system complexity. Complete systems require high bias voltages and usually water cooling.

2.1.3 THz gas laser

THz gas lasers exploit rotational transitions in the excited vibrational state of a low-pressure polar gas molecule. The optically-pumped THz gas lasers rely on the selective absorption of tunable infrared radiation to create a population inversion between rotational states. Fig. 2.1 displays the lasing process: first mid-IR radiation passes through an absorption cell. If the energy of the mid-IR photon matches the transition between the ground and the excited vibrational state, it becomes absorbed by the gas molecules. This process results in a population inversion between rotational states, and an inverted rotational transition causes stimulated emission at the THz frequency. The molecule remains in the excited vibrational state, which must return to the ground state before the next pump cycle.

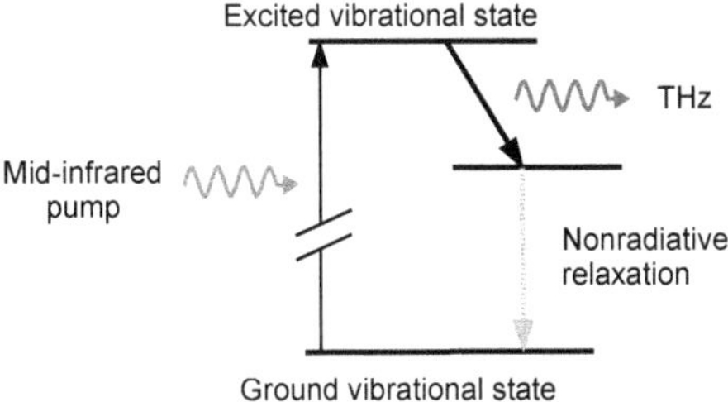

Figure 2.1: Lasing process of a optically excited THz gas laser. Adapted from [23].

For many years, CO_2 pumped gas lasers have been the major source of cw THz radiation above 0.3 THz. Typical CO_2 pumped THz gas lasers cover a frequency range of 0.15–8 THz with output power varying from several μW to several 100 mW [24]. Several groups have recently demonstrated THz radiation in an optically pumped ammonia pumping provided by a mid-IR QCL [25, 26]. Such lasers can provide a THz power level of 20–34 μW for 40–60 mW of pump power [26].

2.1.4 Difference frequency generation

Dual-wavelength mid-IR QCLs are currently the only electrically pumped semiconductor sources operating at room temperature that allows for difference-frequency generation (DFG) within 1–6 THz frequency ranges. These sources rely on the giant susceptibility ($\chi^{(2)}$) in the active region of mid-IR QCLs and a Cherenkov emission scheme [27]. The performance of THz DFG-QCLs has rapidly improved, but still, it can generate an only cw output power of 14 μW [28]. Another device for THz generation is photoconductive antennas (THz-PCAs), which have been extensively used for the generation of THz broadband pulsed and single-frequency cw signals. In cw mode, two laser beams, with their frequency difference in the THz range, combined either inside an optical fiber or overlapped adequately in space, are mixed in a photo-absorbing medium (photomixer) and generate a beat frequency signal. The output of two cw lasers is converted into cw THz radiation exactly at the difference frequency of the lasers. THz signals with the frequency linewidth as low as a few kHz can be generated by photo mixers. The frequency of the THz signal can be tuned by tuning the wavelengths of the lasers. On the other hand, the output power in conventional photomixers falls from 2 μW at 1 THz to below 0.1 μW at 3 THz [22].

2.2 THz detectors

THz detectors can be divided into two broad categories: direct photon detectors and thermal detectors. The operation of photon detectors is based on a photo effect where the radiation is absorbed within the material by interacting directly with bound or free-electrons. The direct photon detectors show a selective wavelength dependence of response per unit incident radiation power. To achieve excellent signal-to-noise performance and very fast response, a photoconductive detector requires cryogenic cooling. This is necessary to prevent the thermal generation of charge carriers. In this thesis work, Ge:Ga photoconductive detectors are commonly used for most of the measurements. On the other hand, in the case of a thermal detector, the incident radiation is absorbed to change the temperature of the corresponding material. The resultant temperature change alters the physical property to generate an electrical signal and does not rely on the photonic nature of the incident radiation. Thermal detectors are generally wavelength-independent, and the signal depends upon the radiant power. The thermal detector can be mainly divided into two categories: room-temperature and helium-cooled detectors. Golay cells, pyroelectric detectors, and microbolometers are room temperature thermal detectors. Highly sensitive detectors such as bolometers in the THz frequency range are mostly helium-cooled.

2.2.1 Ge:Ga photoconductive detectors

In general, THz photon energies are not sufficient to overcome the bandgap energy of an intrinsic semiconductor. In order to detect THz radiation via the photoconductive process, it is necessary to add impurities to the semiconductor. In such an extrinsic semiconductor, a donor or acceptor state exists close to the conduction or valence band so that a low energy photon can excite an electron out of a donor state or a defect electron (hole) out of an acceptor state. An extrinsic photoconductor detector based on Gallium doped Germanium (Ge:Ga) is used to detect THz radiation in this thesis. The detection mechanism is based on a transition from the Ga ground state to the valence band (Fig. 2.2 (a)). Such sensitive photoconductive detectors have been used for many years for low-noise photon detection for wavelengths from 60 to 120 μm [29]. These kinds of devices have high input impedance at low temperatures. The detector mostly used in this thesis has an input impedance of 150 kΩ at 4.2 K.

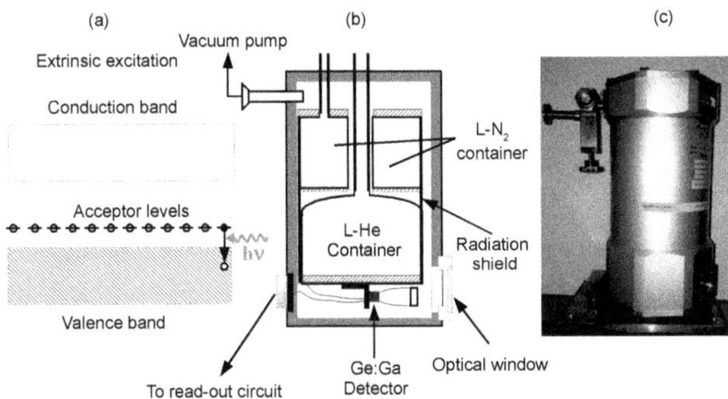

Figure 2.2: a) Extrinsic photoconductor (Ge:Ga) detection mechanism b) Ge:Ga detector mounted inside a liquid helium Dewar c) Photo of a helium Dewar. Figures adapted from [23].

Ge:Ga detectors have to be cooled down to the temperature of the liquid helium to freeze out carriers. As indicated in Fig. 2.2 (b), the sample is mounted on liquid helium cryostat and cooled down to 4.2 K. The base of the liquid helium container is a thick oxygen-free high conductivity copper plate, which has a large thermal conductance. The detector is mounted on a cold copper finger, and a Winston cone is used for condensing the incoming radiation to the detector. A transimpedance amplifier is used as a readout circuit for the detector. Figure 2.2 (c) shows the helium cryostat (HDL 5, Infrared Laboratory) containing two cryogenic vessels. One for liquid nitrogen (LN_2) and one for liquid helium (LHe). The LN_2 vessel directly cools down the radiation shield that covers the work surface, and the LHe cools down the detector to the working temperature. To provide additional shielding, all interior dewar surfaces are covered with metal foil. This cryostat has an LHe capacity of approximately 1.2 liters, and the typical holding time is around 24–48 hours.

2.2.2 Golay cell detectors

Golay cells are room temperature, photo-acoustic devices designed to respond to signals at wavelengths in the range from 20 μm up to a few millimeters. These devices have almost continuous absorption over the entire IR and THz spectrum and have been widely used for THz research for more than 60 years. The mechanism of the Golay detector is shown in Fig. 2.3. At one end, the detector contains a sealed gas cell filled with Xenon due to its low thermal conductivity and, at the other end, a very light flexible mirror. When the radiation passes through a semi-transparent film and is absorbed, the gas heats the pressure inside the gas cell, and the pressure induces a slight movement of the flexible mirror. The optical device shown inside the Golay cell transforms this movement into an electrical signal. A light-emitting diode (LED) emits through a lens system and is then reflected by the flexible mirror back through the grid onto the detector. Any movement in the mirror distorts the reflected image of the grid and changes the amount of light reaching the detector. A chopper is used to allow for the cw radiation to be cut into the AC signal. It reduces the bandwidth and removes the DC drift of the detector. The optimum modulation frequency of these detectors for incoming radiation is within the range of 10–20 Hz. Golay detectors are remarkably sensitive with a typical NEP value of 2×10^{-10} W/$\sqrt{Hz}$ [23].

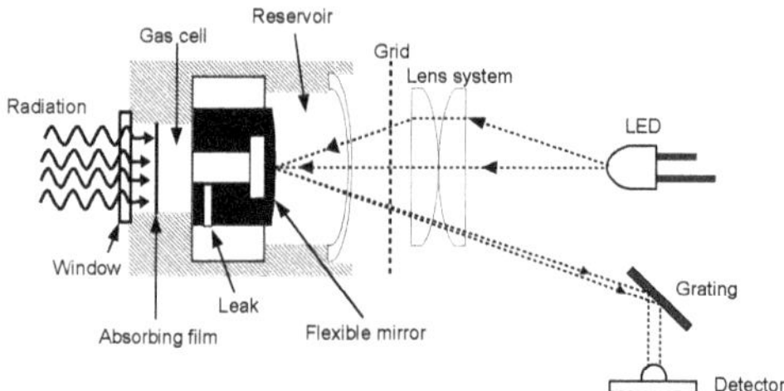

Figure 2.3: Design of a Golay Cell Detector. Adapted from [23].

2.2.3 Microbolometers

Microbolometers are room temperature sensors that are typically used as detectors in thermal cameras. These detectors are usually developed to have the highest sensitivity in a wavelength range of 8–14 μm [30]. Figure 2.4 shows the basic structure of a microbolometer element. The microbolometer consists of a membrane, anchors, and substrate. The membrane is made of a very thin (50–200 nm) amorphous silicon or vanadium oxide (VO_x). It consists of an active sensing surface. When radiation irradiates on the sensing surface of the bolometer, the resulting temperature shift leads to a change of resistance. This change in resistance is analyzed and processed into temperature values, which are then graphically represented in a thermal image. Recently, microbolometer arrays have been developed and optimized specifically for the THz range and are commercially available. Typical video rates today are up to 60 Hz and the NEP is around $10^{-10} - 10^{-11}$ W/$\sqrt{Hz}$ [31].

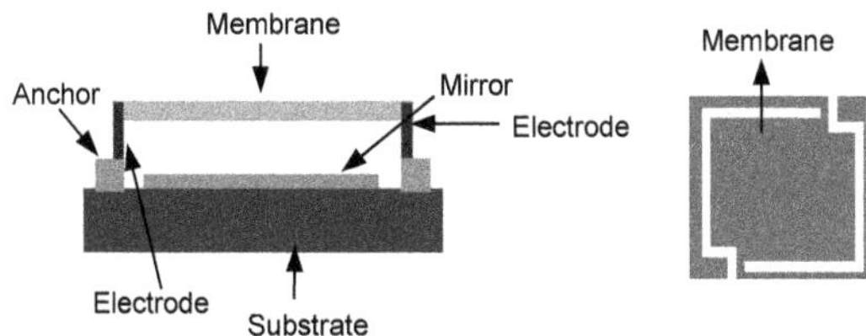

Figure 2.4: Microbolometer structure. Adapted from [30].

2.2.4 Pyroelectric detectors

Pyroelectric detectors are thermal sensors that convert an optical signal into an electrical signal due to temperature changes. Such devices are commercially available either as single detectors or as arrays for the entire IR to THz range. Figure 2.5 illustrates the basic principle of a pyroelectric detector. The pyroelectric detectors are based on polarized crystalline materials which have a permanent electrical dipole moment aligned with a specific crystal axis. When the temperature of the material changes due to optical absorption, the net spontaneous polarization of the pyroelectric crystal changes with a corresponding change of electrical dipole moment as well as the electrical charge. As a result, this electrical charge produces a current flow that can be measured by an external circuit allowing the temperature change to be measured. Pyroelectric detectors are cheap, robust, and, most importantly, work at room temperature. The response time of such devices can be reduced to less than a nanosecond with an appropriate design of the associated amplifier [23]. The typical NEP value for pyroelectric detectors is found around 1×10^{-10} W/$\sqrt{Hz}$ [31].

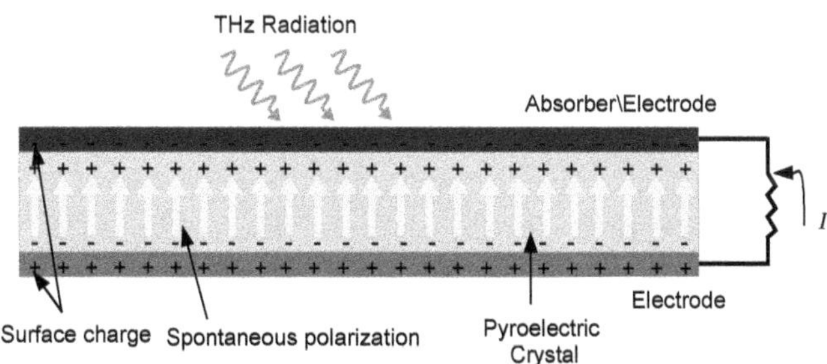

Figure 2.5: Schematic diagram of a pyroelectric detector element. Adapted form [32].

2.3 THz optical components

2.3.1 THz windows

THz QCLs and many sensitive detectors are operated at cryogenic temperatures. As a result, windows are required to hold the vacuum of the cooling system and allow for the THz radiation to be transmitted with as low losses as possible. Due to the high absorption of THz by atmospheric water vapor, the optical pathway often needs to be evacuated, and therefore, windows are also required. The optical windows used in this work and their essential properties will be discussed in this section. Three materials were used mostly as an optical window. These are high-density polyethylene (HDPE), Poly-4-methylpentene-1 (TPX), and crystalline quartz windows.

Table 2.1: Optical constant and suitable frequency range for HDPE, TPX and silica quartz [33, 34, 35].

Material	n	α (cm^{-1})	Frequency range (THz)
HDPE	1.54	0.3	0.29–12
TPX	1.46	0.4	0.1–475
Silica quartz	n_o=2.1, n_e=2.3	4.2	0.29–6

Table 2.1 shows the optical properties of these three materials. HDPE is commonly used as a vacuum seal at cryogenic temperatures because it is chemically inert and mechanically stable. The HDPE refractive index is 1.54, and the coefficient of absorption is 0.3 cm^{-1}, and these values are relatively constant over the whole THz bandwidth. TPX is not only transparent in the THz but also for visible light, and virtually has the same the refractive index ($\sim$1.46) both in the visible and the THz region. The absorption coefficient in the THz region is slightly larger than for HDPE windows. It is more mechanically rigid, but the vacuum seal is slightly lower than for the HDPE material. Crystalline quartz (SiO_2) is a birefringent material commonly used in the manufacture of polarizers (such as waveplates) and windows. The optical losses are high in quartz material as compared to HDPE and TPX because the refractive index and absorption coefficient both are high. The material is birefringent, and incident light splits into ordinary and extraordinary rays. The z-cut is used so that the optic axis of the crystal is perpendicular to the flat surfaces causing the ordinary and extraordinary rays to follow the same path through the crystal and maintain the polarization of the beam.

2.3.2 THz optics

For spectroscopic applications, not only sources and detectors are required but also optical components to guide or manipulate THz waves. This includes mirrors, lenses, polarizers, waveplates, and so on. Metal coated mirrors are used as reflectors for the vast majority of THz systems. Recent developments in THz mirror development include polymeric dielectric mirrors [36], photonic crystal mirrors [37], and semiconductor mirrors [38].

Core components of THz systems include collimating and focusing optics. Off-axis parabolic mirrors are commonly used to collimate THz beams. They are made for a particular angle of incidence of the central portion of the beam, as shown in Fig. 2.6 (a). They are typically coated with materials such as aluminum or gold such that reflectivity of $\geq 99\%$ can be achieved in the THz range. The main advantage of using a parabolic mirror is that it causes fewer losses compared with conventional lenses. It also works over a wide spectral range and is free from spherical and chromatic aberration so that the radiation from the THz source is highly collimated.

Nonetheless, off-axis parabolic mirrors require much space and are very sensitive to misalignment. On the other hand, THz lenses allow for a linear and compact design of the THz device. They are typically manufactured using lathe or milling cutters from polymers material such as HDPE, Teflon, and TPX. Polishing is not needed for many applications that make these lenses relatively inexpensive. Figure 2.6(b) shows the standard TPX lens used for the laboratory experiment. In addition, specialized materials such as picarin [39] and zeoner [40] are available, which, in addition to low losses, exhibit similar refractive index at both THz and visible light frequencies, so that optical pre-alignment is possible.

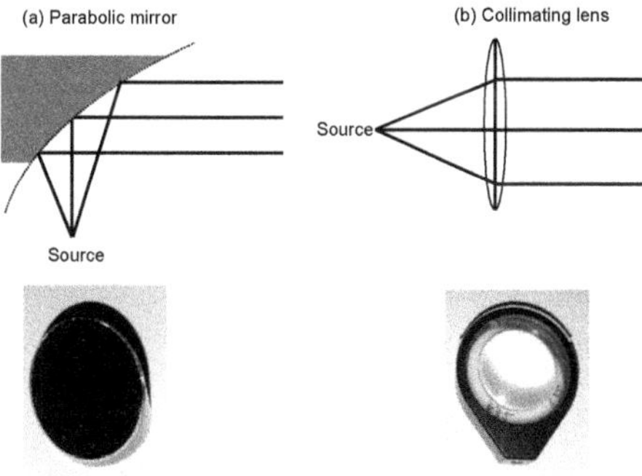

Figure 2.6: a) Off-axis parabolic mirror b) Collimating lens.

2.3.3 Polarizers

In the THz region, free-standing metal wiregrids (WGs) are commonly used as polarizers. They have a high extinction ratio, excellent mechanical durability, and a large acceptance angle for light. Figure 2.7 illustrates the structure of a typical WG polarizer. It consists of thin wires, each of which has a diameter a. The wire grids are positioned in a flat plane and form a regular grid with a spacing g. The properties of wiregrid polarizers are determined by the a, g, and the complex refractive index of the metal. The wires are usually made of tungsten because it has the highest tensile strength among metals and excellent corrosion resistance. A narrow and uniform wire spacing is required to achieve a high extinction ratio in the THz region.

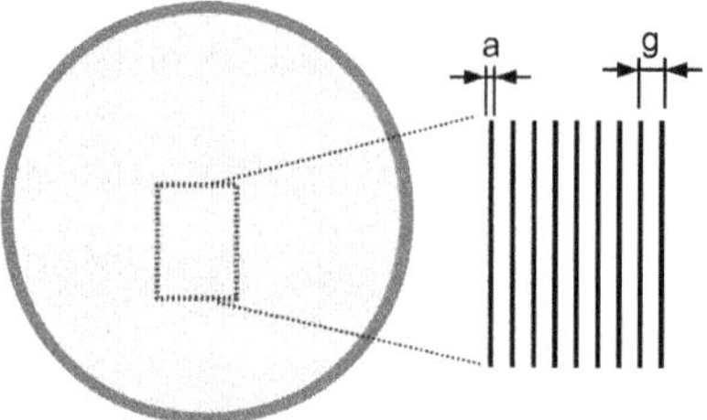

Figure 2.7: Schematic of a WG polarizer. Transmission and extinction ratio are determined by the grid g, and the wire diameter a. Adapted from [32].

The underlying mechanism of such a polarizer is quite simple. If a THz wave impinges on a WG polarizer, and if the electric field is parallel to the wires, the electrons can move freely along the wire direction like a typical metal surface. Thus most of the incident beam is reflected by the polarizer. On the other hand, if the field is perpendicular to the wires, the THz wave does not see much of the wires because the movements of the electrons in the direction perpendicular to the wires are highly restricted. In general, the transmission for an angle θ between the grid direction and the polarization has the following relation: $T(\theta) = \sin^2\theta$. The reflectivity of a WG polarizer is usually higher than 0.95 in a broad spectral range for a field parallel to the grid direction [32]. Thus WG polarizers are often used as beam-splitter.

2.3.4 Waveplates

Waveplates are optical elements that can control the polarization state of light passing through it. Generally, waveplates are made of a birefringent crystal, which exhibits different refractive indices for different polarizations. Using this birefringent property, one can change the phase between the perpendicular polarization components of the incident wave. The most common types of waveplate are half-waveplate ($\lambda/2$) and quarter-waveplate ($\lambda/4$). The half waveplate generates a phase delay of π so that linearly polarized light can be rotated from 0 to 90°. The $\lambda/4$ plate gives a phase delay of $\pi/2$, which changes linearly polarized light to circular and vice versa. Waveplate does not change the polarization of the linearly polarized beam if the polarization direction is along with one of the waveplate axes. A significant disadvantage of the conventional birefringent waveplate is that it can only be used at a single frequency. The waveplates used in this thesis work were made of crystal quartz, which has an excellent birefringent property as well as highly transparent in the THz frequencies [35].

Chapter 3

Fundamentals of terahertz quantum-cascade lasers and laser absorption spectroscopy

This chapter provides a brief description of quantum-cascade lasers and the fundamentals of laser absorption spectroscopy. The chapter begins with the historical overview and the fundamental characteristics of the THz QCLs. The final sections of this chapter focus on the principle and nomenclature of laser absorption spectroscopy and various spectroscopic techniques.

3.1 Historical overview of QCL

QCLs are semiconductor devices that emit in the mid- to far-infrared frequency range (3-200 μm). Such lasers are fundamentally different from the classical diode lasers. The classical diode laser generates light through a photon emission from an electron interband transition. It relies on the recombination of electrons and holes across the bandgap of the used material. However, in the case of QCLs, only electrons are used as charge carriers undergoing transitions between quantized states within the conduction band. The idea of optical gain in the semiconductor superlattice was conceived in the early 1970s. In 1970, Esaki and Tsu published a seminal paper presenting the concept of the minibands and minigaps in periodic semiconductor heterostructures [41]. One year after Esaki and Tsu, Kazarinov and Suris suggested that the optical gain could be achieved by applying an external electrical field to the superlattice structure [42]. However, at that time, QCLs could not be developed due to technical difficulties, i.e., the absence of molecular beam epitaxy (MBE) [43, 44], and metal organic chemical vapor deposition (MOCVD) [44]. It took more than two decades until the first QCL operation was demonstrated with emission at 4.26 μm in 1994 at Bell Labs by Jérôme Faist and co-workers [45]. The performance of QCLs improved drastically in a few years after the first demonstration. Rapid progress was made to obtain operation at room temperature, high optical powers, and emission at additional frequencies. In 1996, first room-temperature operation in pulsed mode was achieved [46], and the wavelength range was extended to 11 μm [47]. However, the first cw operation of a QCL at room temperature was reported only eight years after its first demonstration [48]. Nowadays, mid-IR QCLs have reached impressive levels of performance. Multi-watt output power, cw room temperature (RT) operation has been demonstrated throughout the mid-IR range [44, 49, 50]. The first THz QCL was realized in 2002 by a team led by Alessandro Tredicucci, which lased at 4.4 THz up to 50 K [51]. Although many improvements have been made in terms of output power and emission frequencies since then, the maximum operating temperature is still restricted to values below 210 K [52] in pulsed and 129 K in continuous-wave mode [53].

3.2 Active region of QCL

The active region of a QCL is a semiconductor heterostructure based on tunneling and intersubband transitions. The principle can be illustrated with the help of a 3-level system, as shown in Figs. 3.1(a) and 3.1(b). The arrangement is such that the electrons populate the highest state e_3, then radiate while decaying to state e_2, from which they are extracted to state e_1 that guides them into the next period of the cascade. The injector acts as a ground state for the corresponding period. The role of the injector region is to transport electrons from level e_1 to level e_3 of the next period [54, 55].

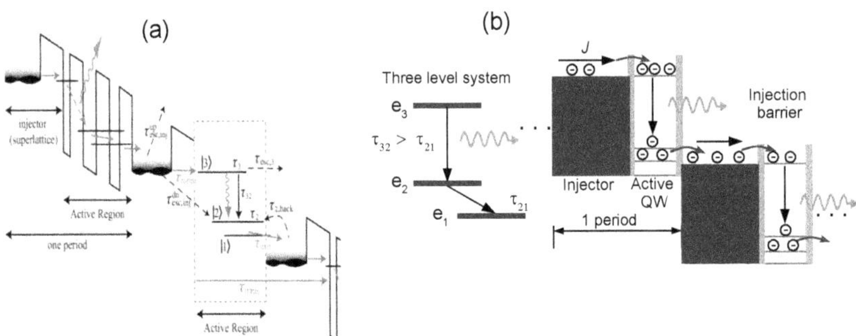

Figure 3.1: Schematic illustration of lasing and cascading in the superlattice structure of the QCL. [55, 56].

Intersubband gain arises from a population inversion between levels e_3 and e_2. Active zone plus injector constitutes the unit of the QCL, called active region period, that is repeated several times during growth to produce the cascade. To achieve population inversion, the non-radiative lifetime τ_{32} of the transition between e_3 and e_2 has to be longer than the lifetime τ_2 of the lower state e_2. For the present discussion, state e_1 is assumed to have a short lifetime and the nonradiative transition $e_3 \rightarrow e_1$ is assumed negligible, hence $\tau_{31} \approx \infty$. An essential parameter for a QCL operation is the injection efficiency. It is defined by the ratio between the injected current to the total current.

$$\eta_i = \frac{J_3}{J}. \tag{3.1}$$

In the ideal case $\eta_i = 1$. Deviation from this value can be due either to thermal activation from the injector to continuum or highly excited states or to direct injection of electrons from the injector to the lower states e_2 and e_1. The second mechanism can be reduced by the use of the narrow quantum well after the injection barrier that enhances the injector coupling with e_3 but lowers the coupling with e_2 and e_1. The first mechanism can be avoided by a proper design of the active zone and by the use of higher energy barriers. If τ_3 is the total lifetime of electrons in level e_3, the steady-state sheet density in this level is given by [56]

$$n_3 = \frac{\eta_i J}{e}\tau_3. \tag{3.2}$$

If one assumes that level e_2 is populated only through the direct scattering of electrons from level e_3, the electron density in e_2 is simply given by

$$n_2 = n_3\frac{\tau_2}{\tau_{32}}. \tag{3.3}$$

where τ_{32} is the mean scattering time from e_3 to e_2 and τ_2 the lifetime of the electrons in e_2. So, the population inversion will be

$$n_3 - n_2 = \frac{\eta_i J}{e}\tau_3(1 - \frac{\tau_2}{\tau_{32}}). \tag{3.4}$$

If all the electrons flow sequentially from e_3 to e_2, n_2 has its maximum value

$$n_2 = \frac{\tau_2 \eta_i J}{e}. \tag{3.5}$$

and

$$n_3 - n_2 = \frac{\eta_i J}{e}(\tau_3 - \tau_2). \tag{3.6}$$

If the condition $\tau_3 > \tau_2$ is satisfied, population inversion and optical gain from the intersubband transition are achieved for any current density. To achieve a low lasing threshold, the QCL should be designed with small optical losses, both by fabricating low-loss waveguides and using long devices to reduce the contribution of mirror loss. To ensure population inversion and hence the dominance of stimulated emission over absorption, the non-radiative scattering mechanisms need to be tailored via the heterostructure design. The frequency of the emission can be tailored since it depends on the energy difference between the involved quantum states.

3.3 Active region designs

The demonstration of THz QCL's active region seemed much more complicated than mid-infrared QCLs. There are mainly two factors that make it challenging to design and fabricate QCLs in the THz range. First, THz photon energy is smaller than the longitudinal optical phonon energy (E_{LO}) of GaAs, which makes it difficult to inject and remove electrons from closely spaced subbands through electron scattering or tunneling. Second, free carrier absorption increases proportionally to the square of the wavelength. It is the dominant factor for waveguide losses in the THz range. These are the limitations always taken into account to design the active region of the QCL described. At this moment, there are four kinds of the active region used in THz QCL: chirped superlattice, bound to continuum, resonant phonon extraction, and hybrid/interlaced design.

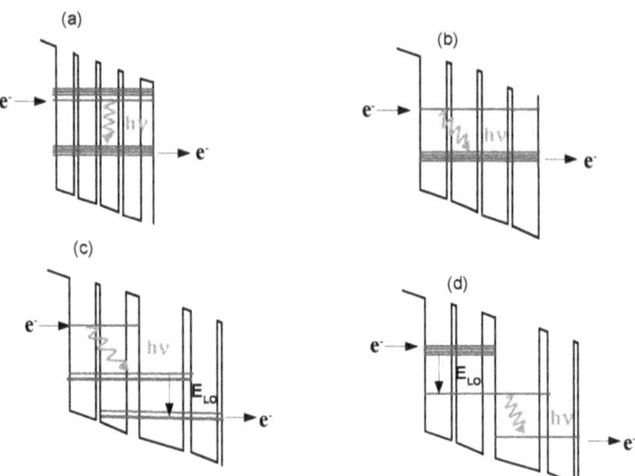

Figure 3.2: Schematic diagram of the active region of THz QCLs based on (a) chirped superlattice, (b) a bound-to-continuum scheme, (c) resonant phonons, and (d)hybrid/interlaced design.

The active region of the first THz QCL was designed based on a chirped superlattice (CSL) (Fig. 3.2(a)). The CSL active region is based on the technique where several quantum wells are coupled together to create minibands. The minibands are designed in such a way that the radiative transition takes place between the lower state of the upper miniband and the upper state of the lower miniband, similar to the conventional interband transition. A population inversion is achieved because the scattering of the electron is tightly confined within the minibands. Figure 3.2(b) demonstrates the BTC scheme in which the radiative transition takes place diagonally between the nonradiative upper state and the upper edge of the lower miniband. THz QCLs with active regions based on CSL and BTC designs exhibit reabsorption of a photon by thermally populated states. To avoid this problem, the resonant phonon scheme has been introduced (Fig. 3.2(c)). In the resonant phonon (RP) system, the injector/collector state is designed to remain at the longitudinal optical phonon energy (36 meV) below the lower radiative state. As a result, the collector state has a fast depopulation by electron-longitudinal optical phonon scattering to depopulate the lower radiative level while maintaining a long upper-level lifetime [57]. The hybrid structure has been developed that employs a resonant-phonon transition for the upper laser population and a miniband for the extraction of carriers from the lower laser level(Fig. 3.2(d)).

The QCL used in this thesis work is mainly based on this hybrid structure. The structure relies on GaAs/$Al_xGa_{1-x}As$ heterostructures with a small Al content of x=0.10–0.15 in the barriers. The smaller Al content allows for the use of thicker barriers to better control the growth of the MBE. The active-region period of the QCLs is based on nine quantum wells. The lasing transition is based on alternating photon and longitudinal optical (LO) phonon-assisted transitions between quasi-minibands [58]. THz QCLs based on the hybrid design exhibit high levels of optical power while the dissipated heat is small enough to achieve cw operation [53].

3.4 Waveguides

The waveguide design and its realization are crucial steps to obtain high-performance QCLs. Mid-IR QCLs use waveguides mostly based on dielectric confinement. However, such a waveguide is impractical in the case of THz QCLs due to strong free carrier absorption at longer wavelengths and the required thick cladding layer. For 1–5 THz radiation, the wavelength of light in the QCL material is between 300 and 60 μm. It is thus very difficult to confine this radiation in the THz QCL active region because its thickness is typically limited to 10 μm, by practical constraints of epitaxial semiconductor growth time [59]. For these reasons, special waveguide techniques have been developed to minimize mode overlap with any doped cladding layers. Currently, there exist two basic types of waveguide for THz QCLs to address this problem: the surface-plasmon (SP), and the metal-metal (MM) waveguide, as shown in Fig. 3.3.

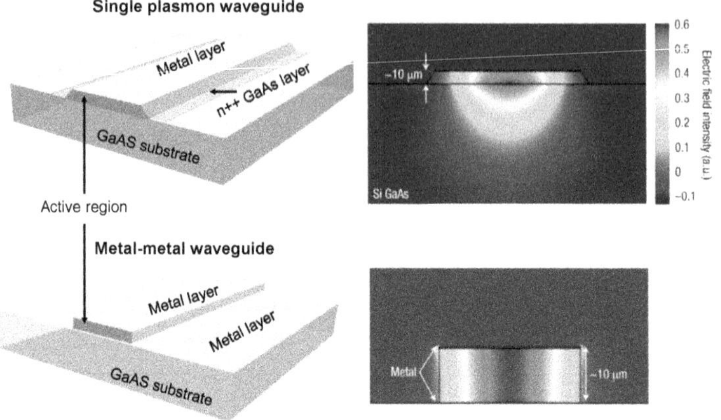

Figure 3.3: Design and mode profile for SP waveguides (top) and MM waveguides (bottom) for THz QCL. Adapted from [57].

In an SP waveguide, the optical mode is confined between a top metal layer and a highly doped GaAs bottom layer. The heavily doped semiconductor layer behaves as a low-density metal. The confined optical mode is partly plasmonic since the plasma frequencies of the metal and the highly doped GaAs layers are well above the frequency of the THz QCL. The larger mode penetration into the substrate also has the advantage that it leads to an emission with good directionality and outcoupling efficiency. In an MM waveguide, the optical mode is sandwiched between two metallic layers so that the optical mode is strongly confined to a sub-wavelength size. Since metals are highly reflective in the THz range, the waveguide mode penetrates only little into the metal layer, and the confinement factor is close to one. In THz QCLs with MM waveguides, 30%–70% of the waveguide losses originates from the losses associated with metal claddings. It has been shown in the literature that optical losses are less by using copper instead of gold cladding [59]. On the other hand, sub-wavelength modal confinement in MM-waveguide also leads to a high facet reflectivity and produce highly divergent far-field emission [60, 57]. For high-temperature operation, MM waveguides are favorable because of their high confinement factor and lower threshold current density [59, 57].

3.5 Temperature performance of THz QCLs

For practical applications of any lasers, the room-temperature operation is always desirable. Mid-IR QCLs operate at room temperature with spectral coverage from 4.5 μm to 20 μm [61, 62, 63]. However, THz QCLs require cryogenic cooling, which is a significant obstacle in making THz QCLs compact sources of radiation. Two major processes are expected to reduce the population inversion of a THz QCL at higher temperatures: thermal backfilling and thermally activated phonon scattering. Backfilling of the lower radiative state with electrons from the densely populated injector occurs either by thermal excitation or by re-absorption of the non-equilibrium longitudinal optical phonons [57, 64]. The thermally activated longitudinal phonon scattering causes the upper-state lifetime to decrease exponentially, which causes population inversion [65]. The first device was designed as a chirped superlattice with a single plasmon waveguide operating up to 50 K in pulsed mode [51]. In general resonant phonon (RP) 3-well achieves better temperature performance than bound to continuum (BTC) design. Well designed metal-metal QCLs with RP active region perform better at a higher temperature. Recently, two groups successfully demonstrated the operation of THz QCL inside a multistage thermoelectric cooler at a radiation frequency of around 3.85 THz and 3.9 THz, respectively [52, 66]. The best device among them exhibits at a record high temperature of 210.5 K (pulsed operation), where the active region of this device is based on a two-quantum-wells period with Cu-Cu waveguide. However, this holds only for pulsed operation [52]. The temperature performance of these devices also depends on the frequency regime in which they are operated. For lower frequencies, the best working temperature in pulsed mode is about 163 K at 1.8 THz [67] and 69 K at 1.2 THz [68].

3.6 Spectroscopic techniques

3.6.1 Laser absorption spectroscopy

For many years, laser absorption spectroscopy (LAS) has been a popular technique for the measurement of gas parameters due to its simplicity, precision, and ability to perform absolute measurements. The main features of the LAS techniques are fast response and higher sensitivity. LAS is often used to exploit the rotational, vibrational absorption characteristics of gas molecules at low moderate pressures to measure the temperature and species concentration. Figure 3.4 shows a typical measurement setup for LAS. The principle of LAS is simple: The laser frequency is swept through a specific range by applying a ramp to the injection current or temperature. If the laser frequency range matches the transition frequency of the atom or molecule of the gas sample, this results in a detector signal similar to that shown in Fig. 3.4 (power vs time). The non-absorbing parts of the ramping frequency signal are used to fit with a baseline, which is used to normalize any changes in the incident laser intensity and enables the measurement of the absolute absorbance using the Beer-Lambert law. Such laser-based spectroscopy enables sensing with high sensitivity. It can provide highly quantitative and selective measurements of several essential species parameters, including gas composition, temperature, and pressure. On the other hand, it relies on the measurement of a small change in intensity and is typically limited to detection of the absorbance $\sim 10^{-3} - 10^{-4}$.

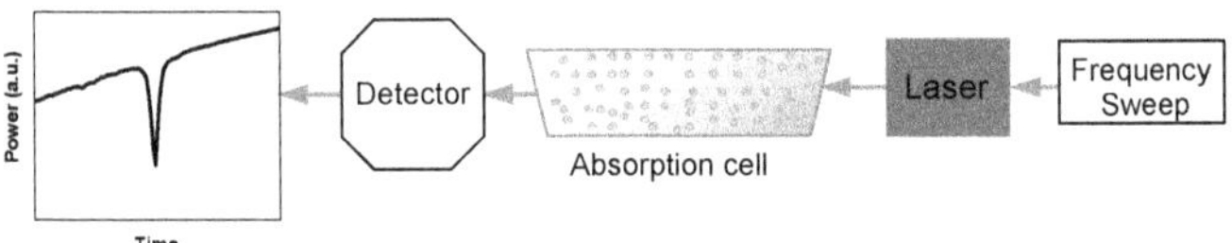

Figure 3.4: Simplified setup of laser absorption spectroscopy.

The Beer-Lambert law

The Beer-Lambert law relates the attenuation of light to the properties of the material through which light is traveling. When a monochromatic optical beam with power P_0 enters an absorption cell, the radiation through the cell with a length L will have the power

$$P(\nu, p, T) = P_0 e^{-\alpha(\nu, p, T)L} \tag{3.7}$$

where $\alpha(\nu, p, T)$ is the frequency-dependent absorption coefficient, and it is most often expressed in cm^{-1}, p is the pressure inside the gas cell, and T is the temperature. According to Beer's law, the absorption coefficient is proportional to the concentration of the absorbing molecules:

$$\alpha(\nu, p, T) = N(p, T)\sigma(\nu, p, T), \tag{3.8}$$

where N is the number density of the absorbing molecules and $\sigma(\nu; p, T)$ is the frequency dependent absorption cross section. Substituting Eq. 3.8 to Eq. 3.7 results in the expression:

$$P(\nu;p,T) = P_0 e^{-N(p,T)\sigma(\nu,p,T)L} \tag{3.9}$$

Eq. 3.7 can also be written as

$$\alpha(\nu;p,T) = -\frac{1}{L}\ln\left[\frac{P(\nu,p,T)}{P_0}\right] \tag{3.10}$$

The absorption coefficient α in Eq. 3.10 can be fitted with a Voigt profile. The broadening effects and their associated line shape will be discussed in the following subsections. The integrated absorption coefficient α_i can be experimentally retrieved from

$$\alpha_i = \int \alpha(\nu,P,T)d\nu \tag{3.11}$$

The integrated absorption coefficient can also be obtained from a molecular database such as the Jet Propulsion Laboratory (JPL) catalog.

Spectral line broadening

Spectral lines in discrete absorption spectra are never strictly monochromatic. There are several reasons for spectral line broadening. One of them is a quantum mechanical reason. Every excited state has an intrinsic lifetime before it spontaneously decays to lower energy levels. This gives rise to the Lorentzian absorption cross-section. The natural line broadening is a result of Heisenberg's uncertainty principle, which indicates that there is an uncertainty in the energy state, ΔE of a system, and due to uncertainty in the lifetime. If the system survives in a quantum state for a time Δt, the energy of the level in principle cannot be known with accuracy better than spontaneous decay [69],

$$\Delta E \Delta t \geq \frac{\hbar}{2} = \frac{h}{4\pi}. \tag{3.12}$$

The spectral distribution of intensity $I(\nu)$ can be written as

$$I(\nu) = \frac{\gamma_L}{2\pi[(\nu-\nu_0)^2+(\gamma_L/2)^2]}, \tag{3.13}$$

where ν_0 is the frequency of the transition in an unperturbed state and $\Delta\nu = \gamma_L/2\pi$ is the full-width-at-half-maximum (FWHM). γ_L is the damping factor defined as $\tau = 1/\gamma_L$. The natural linewidth can be written

$$\Delta\nu = \frac{A_i}{2\pi} = \frac{1}{2\pi\tau} \tag{3.14}$$

where τ is the mean lifetime of the initial state and A_i is the total transition probability.

Pressure broadening

Pressure broadening of the spectral line is caused by collisions of neighboring atoms or molecules. For a high-pressure gas, the collision of other particles with the emitting light particle interrupts the emission process. The duration of the collision is much shorter than the lifetime of the emission process. This disruption plays a vital role in the formation of spectral lines, which follows a Lorentzian line shape. When an atom A with energy level E_i and E_k approaches another atom B, the mutual interaction between A and B can be radiative or nonradiative. If no internal energy of the collision partners is transferred during the collision by nonradiative transitions, the collision is termed as elastic. The line shift caused by elastic collisions corresponds to an energy shift of $\Delta E = \hbar\Delta\omega$. Besides elastic collisions, inelastic collisions may also occur in which the excitation energy E_i of atom A is either partly or entirely transferred into internal energy of the collision partner B or translational energy of both partners. The total transition probability A_i for the depopulation of level E_i is a sum of radiative A_i^{rad} and collision-induced A_i^{coll} between the mean relative velocity υ, the responsible pressure p, and the gas temperature T gives the total transition probability [69].

$$A_i = A_i^{rad} + A_i^{coll} \quad with \quad A_i^{coll} = N_B \sigma_i \upsilon \tag{3.15}$$

where N_B is the number density of collision partners, and σ_i is the cross section of atom.

$$\upsilon = \sqrt{\frac{8kT}{\pi\mu}}, \quad \mu = \frac{M_A \cdot M_B}{M_A + M_B}, \quad p = N_B kT \tag{3.16}$$

k the Boltzmann constant, M_A and M_B the mass of the atoms A and B, respectively.

$$A_i = \frac{1}{\tau_{sp}} + b\,p \quad with \quad b = 2\sigma_i \sqrt{\frac{2}{\pi\mu kT}} \tag{3.17}$$

It can thus be seen from Eq. 3.14 and 3.17 that the pressure broadening $\Delta\nu_p$ can be described:

$$\Delta\nu_p \approx b\,p \tag{3.18}$$

The pressure broadening $\Delta\nu_p$ (FWHM) can be expressed in terms of the pressure broadening coefficient b and the gas pressure p.

Doppler broadening

The thermal motion of molecules shifts the apparent frequency of the emitter as a result of the Doppler effect. The Doppler effect results in a light frequency shift when the source is moving toward or away from the observer. It is an inhomogeneous effect as the probability of absorption transition is not equal for all molecules of the same gas species. It depends on the velocity component class of the random motion of molecules described by a Maxwell velocity distribution function, the mass of the emitter, the frequency of the line, and the temperature. When an atom or molecule is traveling with a velocity **v**, then the frequency ν_a at which a transition is observed to occur is related to the transition frequency ν in a stationary atom or molecule by [69]

$$\nu_a = \nu_0\left(1 \pm \frac{V_z}{c}\right)^{-1}, \tag{3.19}$$

where c is the speed of light and the sign $+$ and $-$ is related to an approaching or receding source, respectively. Because of the usual Maxwell velocity distribution, there is a spread of values of the velocity component V_z and the shape of pure Doppler broadened absorption lines are Gaussian, with an FWHM given by

$$\Delta\nu_D(FWHM) = 2\frac{\nu_0}{c}\left(\frac{2kT\ln 2}{m}\right)^{1/2} = 7.17\times 10^{-7}\nu_0\sqrt{\frac{T}{M}}, \tag{3.20}$$

where M [g-mol^{-1}] is the molecular weight of the absorbing molecule. From the above equation, it is seen that $\Delta\nu_D$ is a function of the transition line center frequency ν_0, the molecular weight of the absorbing species m, and the temperature T.

Combined broadening

As discussed above, in the case of pressure and Doppler broadened lines, the lineshape can be approximated by Gaussian and Lorentzian lineshapes, respectively. In the intermediate case ($\Delta\nu_D \approx \Delta\nu_p$), where neither Doppler nor collisional broadening is the dominating mechanism, the lineshape function takes the form of a Voigt profile, which is the convolution of Gaussian and Lorentzian lineshape function. An expression for the normalized Voigt profile is given by [70]

$$I(\nu) = \frac{2\Delta\nu_p}{\Delta\nu_D^2}\frac{1}{\pi^{3/2}}\int_{-\infty}^{+\infty}\frac{e^{-y^2}}{\left(\frac{2(\nu-\nu_0)}{\Delta\nu_D}-y\right)^2+\left(\frac{\Delta\nu_p}{\Delta\nu_D}\right)^2}dy \tag{3.21}$$

For Lorentz and Doppler profiles with the same FWHMs, the corresponding Voigt profile is steeper than the Lorentz profile and flatter than the Doppler profile in the line center.

3.6.2 Heterodyne spectroscopy

Heterodyne detection for astronomical and Earth observations in the THz region is a reliable and significant spectroscopic technique. A schematic of heterodyne detection is shown in Fig. 3.5. The detection method is based on frequency down-conversion in a non-linear device, accomplished by mixing the signal with a reference signal at the fixed frequency. The basic principles of this technique are explained below.

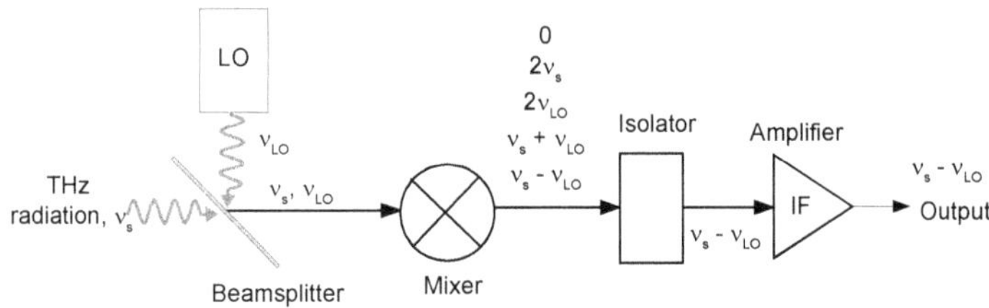

Figure 3.5: Schematic setup for a heterodyne receiver.

In this detection scheme, the signal of interest is mixed with a reference signal from a local oscillator (LO). The LO has to provide a narrow, single-mode frequency to drive the mixing process. Both frequencies from the signal of interest and the reference signal from the LO are mixed in a nonlinear optical element called a mixer. The resulting mixer signal is proportional the total optical intensity I [71]:

$$
\begin{aligned}
I &\propto [E_s \cos(\nu_s t + \phi) + E_{LO} \cos(\nu_{LO} t)]^2 \\
&\propto E_s^2 (1 + \cos(2\nu_s t + 2\phi)) + E_{LO}^2 (1 + \cos(2\nu_{LO} t)) \\
&+ E_s E_{LO} [\cos((\nu_s + \nu_{LO}) t + \phi) + \cos((\nu_s - \nu_{LO}) t + \phi)] + ..
\end{aligned} \tag{3.22}
$$

The output signal has high-frequency components ($2\nu_s$ and $2\nu_{LO}$) and two mixed products with the sum frequency ($\nu_s + \nu_{LO}$) or the difference frequency ($\nu_s - \nu_{LO}$) component. The intensity of the difference frequency term is of significance. It carries all relevant information of the original signal (amplitude, phase) at much lower frequencies that can be further processed by electronics (e.g., using amplifiers, filters). Heterodyne detection offers high spectral resolution $\nu/\Delta\nu \sim 10^5 - 10^6$. It can detect both intensity and phase. The feature of this detection technique is that it has less background noise compares to the direct detection method, and noise mostly depends on the fluctuation of the local oscillator. There are some difficulties with heterodyne receivers: both the LO and THz beam should be coincident, and their Poynting vector should be coincident [72]. In the THz range, three types of mixers are often used: Schottky-barrier diodes [73], superconductor-insulator-superconductor junctions (SIS) [74] and superconducting hot electron-bolometers (HEB) [75]. Schottky diode mixers are commonly used in the spectral range below 1 THz. They have low sensitivity and require several milliwatts of LO power. The SIS mixer is also used for low frequencies. They have a moderate bandwidth and can reach quantum-limited noise levels. HEBs are the most sensitive mixers for high-frequency (> 1.4 THz) applications. The common type of HEBs is an NbN bolometer with high sensitivity and short response time in a broad spectral range.

3.6.3 Modulation spectroscopy

Modulation spectroscopy is a high-resolution spectroscopic technique capable of sensitively detecting spectral absorption and dispersion features. It provides a high signal-to-noise ratio of the detected signal at a fast detection speed. Modulation spectroscopy with semiconductor lasers can be performed by modulating the injection current [76, 77] so that no external electro-optical modulators are required [78]. A general way to see modulation spectroscopy is as a sinusoidal modulation of the phase of the electromagnetic field. The field generated by a laser radiating at an optical frequency ν_0 with sinusoidal phase modulation is described by [76]

$$E(t) = E_0 \exp[i(2\pi\nu_0 t + \beta \sin(2\pi\nu_m t))], \tag{3.23}$$

where E_0 is the laser field, ν_m the modulation frequency, and β the phase modulation amplitude. Methods that use modulation frequencies higher than full-width at half-maximum (FWHM) of the absorption line (typically 50 MHz to several GHz) in combination with small modulation indices ($\beta \leq 1$) are referred to as frequency modulation spectroscopy (FMS). On the other hand, techniques that make use of large modulation indices ($\beta \gg 1$) and modulation frequencies that are small as compared to the FWHM of the absorption line are called wavelength modulation spectroscopy (WMS). The high-frequency WMS ($\geq$100 kHz) offers excellent sensitivity without the burden of extremely fast detection electronics, as required by FMS.

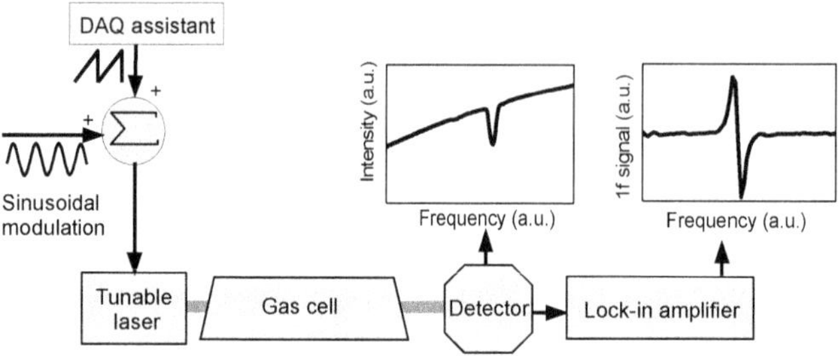

Figure 3.6: Schematic setup for a frequency modulation spectroscopy. Left panel: direct signal, right panel: 1f signal.

Figure 3.6 shows a typical setup for modulation spectroscopy. Like standard direct-absorption measurements, the laser frequency is tuned by ramping up the laser injection current. An additional sinusoidal signal is superimposed on the injection-current in order to generate additional high-frequency modulation in both laser intensity and frequency. When the laser frequency is tuned across the absorption line, the frequency modulated signal causes the laser to be tuned back and forth across the absorption line. As a result, absorption affects the shape of the transmitted laser intensity and introduces harmonic components to the detected signal. Subsequent demodulation of these harmonics at a narrow frequency interval is performed by a lock-in amplifier, which results in significantly better signal contrast. These highly-sensitive measurement techniques have been increasingly applied for measurements in harsh environments due to the improved sensitivity and noise-rejection capabilities over direct absorption [79]. Modulation spectroscopy is also often applied for laser frequency stabilization to a molecular absorption line.

3.6.4 Saturation spectroscopy

Saturation spectroscopy is a high-resolution spectroscopic technique capable of detecting the rotational transition of molecular lines free of Doppler broadening. This technique has often been used to study the hyperfine and isotopic structure of atomic or molecular absorption lines [80]. Simplified setup for saturation spectroscopy is shown in Fig. 3.7.

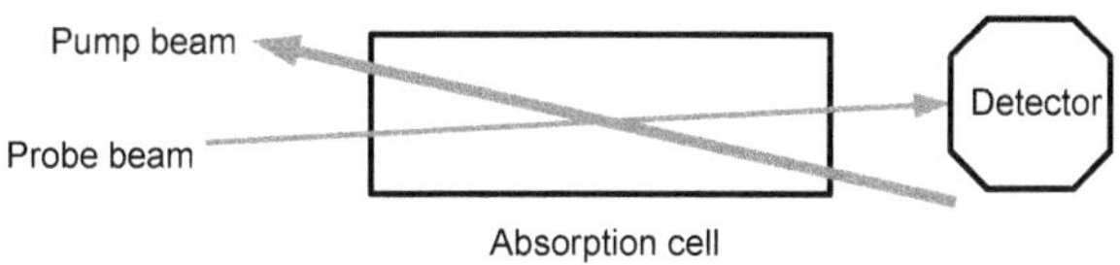

Figure 3.7: Schematic setup for saturation spectroscopy.

For this measurement, a relatively high-intensity laser pump beam is sent through the absorption cell, and a weak probe beam is also sent from the same laser to the opposite direction of the absorption cell and the detector. The strong pump beam saturates a transition, and the same transition is then probed by that counter-propagating weaker beam. There are some prerequisites for saturation spectroscopy that include (i) very low pressure inside the absorption cell ($\leq 10^{-3}$ hPa) to avoid pressure broadening (ii) high pump power to saturate the molecular transition, and (iii) narrow laser linewidth to resolve the Doppler-free feature. Details of this method will be discussed in chapter 5.

Chapter 4

Experimental techniques

This chapter will deal with the technical setup and experimental methods used to obtain the results presented in this work. Section 4.1 introduces the cooling systems employed to operate the THz QCLs to temperatures usually well below 70 K. Sections 4.3–4.4 comprise the basic THz QCL characterization techniques that are crucial for this work. Section 4.5, provides an introduction to Fourier transform spectrometer, which is used to determine the emission spectra of THz QCLs. Finally, the setup, data acquisition, and calibration for high-resolution molecular spectroscopy will be explained in section 4.6.

4.1 Cooling system

As discussed in the previous chapter, THz QCLs are operated at cryogenic temperatures. For a decent performance of THz QCLs, a cooling system is required, which allows for operation below the boiling temperature of the liquid nitrogen. Two cooling systems are used to study THz QCLs: a liquid helium flow cryostat and a mechanical cryocooler. Details of these two cooling systems and their technical configuration for THz QCLs will be discussed in the following.

4.1.1 Helium flow cryostat

The liquid helium flow cryostat (HFC) (Oxford Instruments, Optistat CF-V) used in this thesis can cool the sample down to 4.2 K. Fig. 4.1 shows the configuration of the HFC for THz QCLs. A customized cold finger is used for the QCL mount. The thermal insulation is realized by evacuating the chamber in which the cold finger and the sample are located. The vacuum is retained inside the HFC by means of a turbopump. The cryostat is continuously refilled from a storage tank using a transfer tube with a flow rate of typically less than 0.5 l/hour. A control valve is used to regulate the flow of helium.

The QCL is mounted in such a way that the heat sink is always in good thermal contact with the cold finger. A differentially cooled radiation shield is used to keep the absorption of heat radiation to a minimum. The temperature sensor (Cernox, CX-1050-AA-1.4L) and the resistive heater are incorporated into the cold finger of the HFC to stabilize the temperature with the PID control loop using a temperature controller (Cryocon, 24C). A TPX (poly-4-methyl pentene-1) window with a diameter of 42 mm is used to obtain optical access to the QCL. HFCs are commonly used in low-temperature experiments due to their easy operation and fast cooling, especially when mechanical vibration prevents the use of cryocoolers.

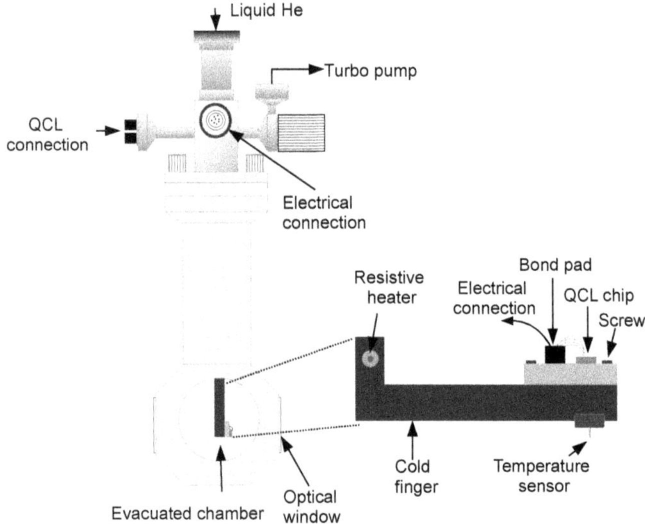

Figure 4.1: THz QCL inside a He-flow cryostat.

4.1.2 Mechanical cryocooler

Mechanical cryocoolers are mainly used for practical applications such as space science missions because the need for liquid helium in HFCs restricts their applicability to laboratories. A twin-piston linear integral mechanical cryo-cooler (Ricor, K535) is used to cool down the QCL. This dry cooling technology offers the possibility to cool down the sample without any cryogenic fluid. Figure 4.2(a) shows a photo of the mechanical cryocooler used to study THz QCL for this thesis. It is a tabletop cooling device that allows easy access to the sample and can cool the sample down to 35 K. The cooler unit weighs approximately 9.5 kg, is approximately 40 cm long and draws less than 400 W of electrical power with a cooling capacity of 1.7 W at 45 K. The power supply for the cooler unit weighs 2.6 kg and has the dimensions $10.2 \times 13.0 \times 33.6\ cm^3$. The submount with the QCL is attached to a second copper mount, which in turn is fixed to the cold finger of the Stirling cooler. In order to ensure tight thermal contacts, indium foil is placed between the two copper mounts as well as between the second copper mount and the cold finger. Figures 4.2(b) and 4.2(c) show both the schematic and the photograph of how the QCL is mounted on a cold finger inside a mechanical cryocooler. The operation of mechanical cryocoolers can be explained in a simplified manner as follows: it consists of three main components: the expander, the compressor and the thermal shroud. The cold finger on which the QCL is mounted in front of the expander. The expander is linked to a compressor which provides the required helium gas flow rate at high and low pressure to cool down the expander to the required temperature. The thermal shroud allows the heating/cooling of items in the chamber under vacuum. For the QCL characterization, thermal stabilization is achieved by means of a temperature sensor and a heater operated by a temperature controller. The primary drawbacks of the dry cooling method are that they produce unavoidable vibrations.

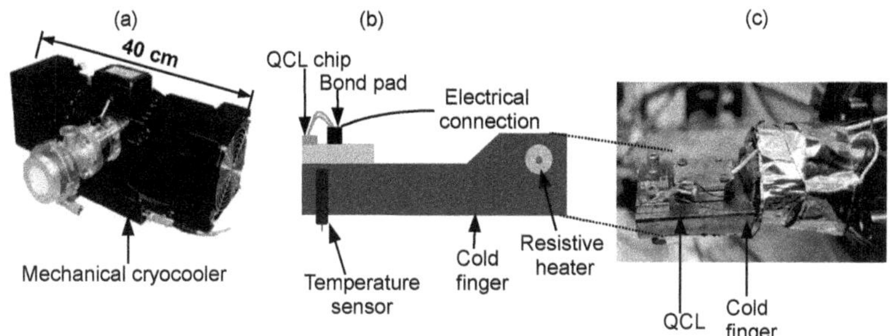

Figure 4.2: THz QCL mounting inside a mechanical cryocooler.

4.2 Power measurement of THz QCLs

Precise measurement of THz power is necessary for a variety of technical applications, including THz imaging and spectroscopy. Only a few commercial THz detectors are on the market that can reliably measure the absolute power of THz radiation. A Thomas Keating (TK) instrument measurement system (TK-THz-PM07) is used in this experiment to determine absolute THz power. The typical setup for the power measurement is shown in Fig. 4.3. The power meter head consists of two closely spaced parallel windows which form an air-filled cell. A thin metal film is inserted between the gap of the windows, which absorbs a known fractional of incident THz radiation. When the beam from the QCL impinges on the film, a fraction of the power is absorbed by the film. The QCL radiation is chopped as a square wave by an optical chopper. In turn, this produces a modulation in the thin film's temperature resulting in a modulation of the pressure in the cell. This modulation is detected by a pressure transducer and measured with a lock-in amplifier and displayed on the PC. The modulated pressure change is closely proportional to the total absorbed power; it depends negligibly on the distribution of the absorbed power across the film. The frequency of the chopper is in the range 10 Hz to 50 Hz, and the lowest noise equivalent power (NEP) is found between 20–30 Hz. The power is measured as a function of the applied current to the THz QCL. The power meter head should be aligned in a Brewster angle ($55.5^{\circ} \pm 5^{\circ}$) to avoid reflection from the window. It can be calibrated by dissipating a measured amount of ohmic power in the film. The maximum power that can be measured with this detector without saturation is around 200 mW. The system has a typical NEP of about 30 $\mu W/\sqrt{Hz}$ [81].

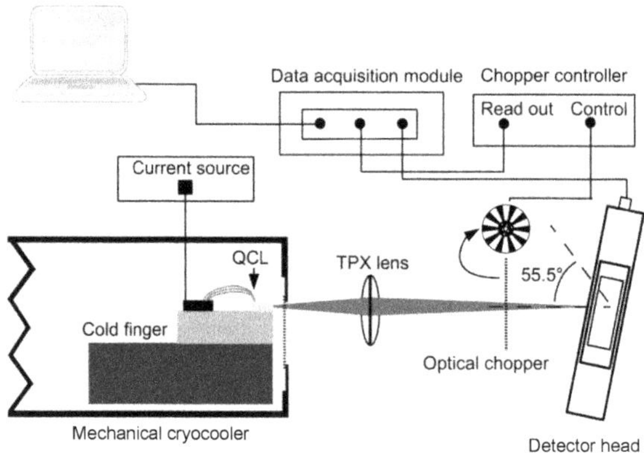

Figure 4.3: Typical absolute power measurement setup for THz QCL.

4.3 Light-current-voltage characterization

The light-current-voltage (LIV) sweep characterization is the basis for spectroscopic techniques. Fig. 4.4 shows an overview of the characterization setup and the corresponding electric circuit and an example for LIV characteristics. The optical intensity from the QCL is collected by a TPX lens and then directed into a fast photo-conductive He-cooled Ge:Ga detector, which is positioned in front of the laser, as can be seen from the schematics. A fast detector is used to keep the measurement time as short as possible in order to prevent self-heating. The drive current is swept from below threshold up to a maximum level of the specific QCL, and the detector voltage is recorded. The voltage output from the Ge:Ga detector is then calibrated with the Thomas Keating (TK-THz-PM07) power meter to find the absolute power. These measurements are repeated at different temperatures to deduce the thermal behavior of the device. In Fig. 4.4(b), the DC current ramp is applied via a CompactDAQ-system from national instruments (NI), and these current ramps are applied to the QCL via a low impedance line. The bias voltage can be measured as the voltage drop over the laser, that is, the potential difference between the two electrodes of the QCL. Fig. 4.4(c) shows typical LIV characteristics of a QCL at different temperatures. The current-voltage (IV) characteristics show the initial increase of the applied bias voltage without any current flowing through the quantum well structure of the active region. As expected, with increasing temperature, the threshold current density increases and the output power decreases.

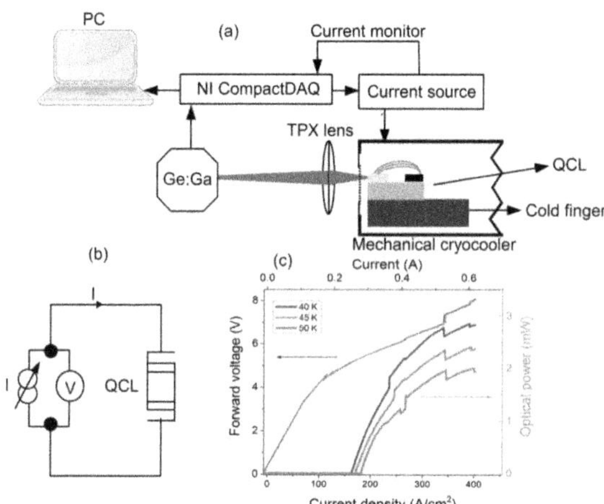

Figure 4.4: (a) Schematic configuration for light-current-voltage (LIV) characteristics (b) the DC sweep circuit (c) LIV characteristics of THz QCL.

4.4 Beam characterization

The characterization of the QCL beam profile is one of the most important and fundamental measurements for many QCL based applications. Most applications require the power available in the fundamental Gaussian mode. The beam profile of a THz QCL depends strongly on the type of waveguide. For this work, the QCLs used are based on the SP waveguide. The beam profile of a QCL with an SP waveguide has contributions from both the laser ridge and the substrate. With such a laser, an M^2 value of less than 1.2 has been demonstrated [82]. Figure 4.5 shows the schematic configuration for a beam profile measurement. The QCL that is used for this experiment emits a single mode at 3.3 THz. It is operated inside a mechanical cryocooler. The employed camera (Automation Technology, IRS–324) is equipped with a microbolometer array which has a matrix of 324×256 pixels with a pixel pitch of 25 μm. To improve the sensitivity of the microbolometer camera at THz frequencies, the germanium objective lens in front of the camera is removed and replaced with a TPX lens. As can be seen from Fig. 4.5, a 25 mm biconvex lens has been used between the camera and the QCL to measure the beam spot. The camera is positioned exactly 25 mm away from the QCL. The radiation from the QCL is aligned, such to obtain the beam spot just at the center of the camera sensor. The camera is connected to the computer to monitor the quality and record the real-time image of the QCL beam.

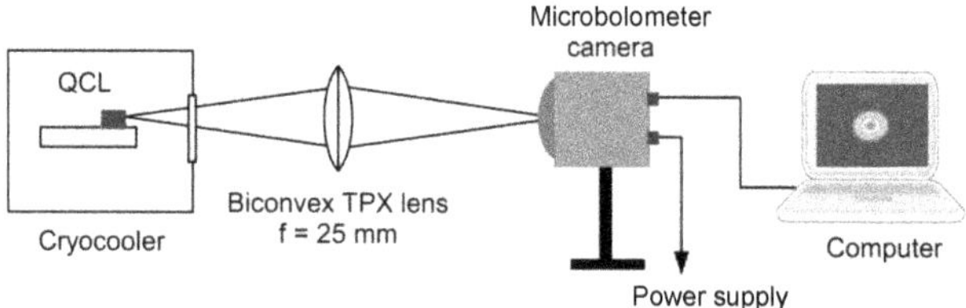

Figure 4.5: Experimental setup for beam profile characterization.

Figure 4.6 (a) shows a typical QCL beam profile measured with a microbolometer camera. The red and blue Gaussian curve in Fig. 4.6 (b) and 4.6 (c) display a horizontal and vertical cross-section, respectively of the beam profile, corresponding to the profile shown in Fig. 4.6 (a). The dotted lines refer to the measurement data, and the solid lines are the result of a Gaussian fit. The beam profile of the QCL is almost Gaussian-shaped, except for some small side lobes. The side lobes are caused by diffraction at the TPX lens. From the Gaussian fit (FWHM), the beam waist of this particular measure is determined around 0.23 mm in horizontal and vertical directions.

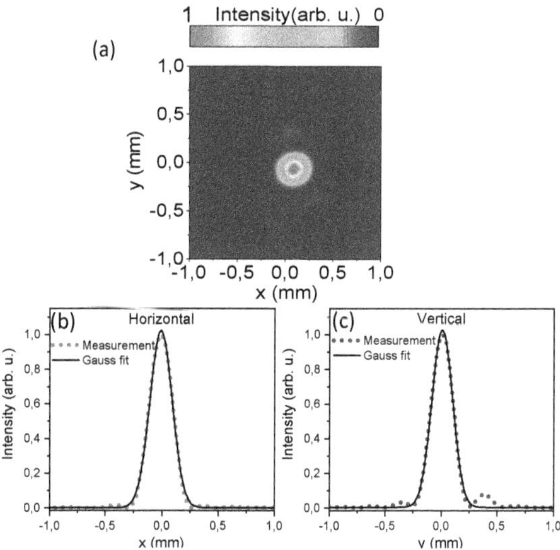

Figure 4.6: (a) Beam profile at the center position ($x = y = 0$) of the microbolometer camera. (b) Horizontal cross section, and (c) Vertical cross section of the beam profile.

4.5 Fourier transform spectroscopy

Fourier transform spectroscopy is a well-known technique for measuring the emission spectrum of the radiation source or the absorption spectrum of a sample. Figure 4.7 (a) shows the measurement setup of a Fourier transform spectrometer (FTS) with a THz QCL. The incident source is collimated by a 90° off-axis parabolic mirror and then split into two parallel beams by a Mylar beam splitter. After being reflected by the two mirrors, the beams are recombined with the same beam splitter and sent to the Golay cell and detected by a lock-in amplifier. A chopper is used in front of the detector to modulate the radiation. To measure the transmission spectrum, one of the interferometer mirrors is moved continuously so that the intensity of the radiation impinging on the detector is modulated periodically due to interference. The detector signal as a function of the path length difference between the two mirrors is usually referred to as interferogram. The corresponding spectrum is obtained through a numerical Fourier transformation of the interferogram performed by a LabVIEW routine, and its resolution is determined by the maximum path length difference. Figure 4.7 (b) shows the measured spectrum of a 3.3 THz QCL. The horizontal axis is the frequency, and the vertical axis is the normalized intensity.

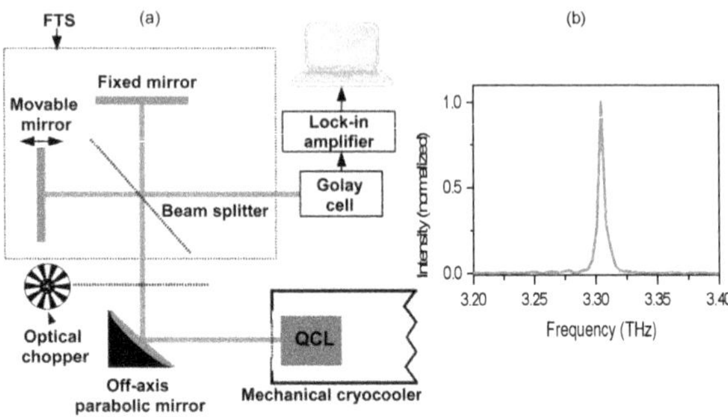

Figure 4.7: (a) Simplified setup of a QCL based Fourier Transform spectrometer, and (b) spectrum of a 3.3 THz QCL.

For astronomical observation, the required spectral resolution is mostly determined by the Doppler width, which is typically a few MHz at a transition frequency of 1 THz [83]. The spectral resolution must be in the order of $\Delta\nu/\nu = 10^{-6}$ to conduct this kind of high-resolution spectroscopy. Although, FTS can provide adequate frequency coverage, but can not provide resolution at this level. In addition, broadband THz FTS instruments having large bandwidth leads to a 'multiplex disadvantage' because the room-temperature laboratory background can easily saturate the sensitive THz detectors. Other sensitive techniques, such as heterodyne spectroscopy, are typically used for this type of measurement.

4.6 High-resolution molecular spectroscopy with a THz QCL

High-resolution molecular spectroscopy is an appropriate tool to analyze the structure and energy levels of molecules and atoms. The spectroscopy in the THz range is important because of the so-called fingerprint of many species in that range. Performing molecular spectroscopy in the THz range allows for appealing fundamental research in physics [84, 85] and also has a significant impact on environmental monitoring and space research [86, 87]. The invention of QCLs was a crucial step towards the advancement of high-resolution spectroscopy. THz QCL-based spectrometers are particularly attractive because they can provide high cw output power up to a few hundred mW [88], narrow linewidth [89] and sufficient frequency tunability of several GHz. In the following section, a QCL-based measurement system for high-resolution spectroscopy, the corresponding data acquisition technique, and a frequency calibration method will be discussed in more detail.

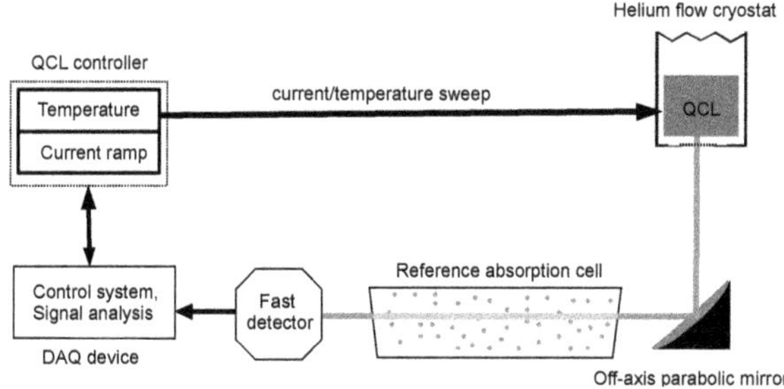

Figure 4.8: Schematic of the experimental setup for QCL-based spectrometer.

4.6.1 Measurement setup

The experimental method used in this thesis involves a narrow-linewidth tunable QCL (a single-mode laser), which is tuned across a specific frequency range. The absorption of light is analyzed as a function of that frequency. Figure 4.8 shows the usual experimental setup, which is similar to Fig. 4.4, but in this case, the absorption cell is used between the QCL and the detector. The spectroscopic setup consists mainly of an optical unit and an electronic unit. The optical unit is often built on an optical table and includes QCLs and detectors, moving optics, and a reference absorption cell. The electronics unit consists of a data analysis computer, a data acquisition module, and a specially designed laser control module.

The THz QCLs are housed in a liquid helium-cooled dewar or a mechanical cryocooler. For optical collimation, either a TPX lens or a parabolic mirror is used to direct the emission from the QCL to the absorption cell and the detector. A Ge:Ga photoconductive detector is employed because the detector is fast and sensitive in the THz frequency range. For gas spectroscopy, a 60 cm long single-pass absorption cell has been used for all measurements presented in this thesis. The typical gas cell consists of a silica cylinder mounted in aluminum receptacles with tilted HDPE windows. The sample gas is injected at constant flow through a needle valve at one end of the gas cell. Another needle valve is used, which is attached to the other end and connected to the turbo-molecular pump for evacuation. The windows are tilted due to avoid any standing waves. Before filling with a sample gas, the absorption cell was purged by nitrogen to prevent any contamination. A pressure gauge is used to monitor the gas pressure inside the gas cell. For calibration, CH_3OH is used as a molecular species because its absorption spectrum is well documented. The frequency of the QCL is tuned across a specific frequency range, either by changing its driving current or its heat-sink temperature, and the corresponding spectra are typically measured at a pressure of 0.1–1 hPa. Another fast tuning technique that is developed to sweep the QCL frequency is based on illuminating the back facet of the QCL, which will be explained in the later chapter. The driving current for the QCL is typically supplied by a low noise current source to reduce the intensity and frequency noise. The results obtained with a 3.1 THz QCL are exemplarily discussed in the following.

4.6.2 Data Acquisition

One of the challenges in experimental work is to define the best way to record and store the data generated by the experiment. This requires not only coordinating hardware (time, synchronization, speeds, etc.) among different acquisitions but also choosing the relevant parameter to speed up data analysis without compromising essential features. In this section, we start by explaining the data acquisition used in our spectroscopic system. A fast data acquisition device (CDAQ, National Instruments) is used to obtain data from the measurement. A LabVIEW routine is developed to tune the QCLs, to record the signal from the detectors, and to save the information. In order to characterize the full range of the emission frequency of the QCL, a fast sweep (20–80 ms) is applied to the driving current of the QCL from below the threshold to the maximum for the various temperatures of the heat sink between 40 and 70 K with a step size of approximately 0.1 K. Figure 4.9 (a) shows the signal transmitted through the absorption cell which is filled with CH_3OH as a function of driving current and heat sink temperature of the QCL. The corresponding transmission signal at a fixed point in the map (41 K and 350 mA) is shown in Figs. 4.9 (b) and 4.9 (c).

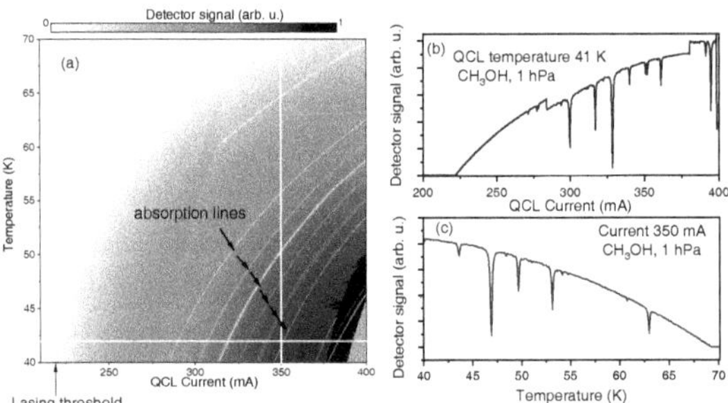

Figure 4.9: (a) Signal transmitted through the absorption cell filled with CH_3OH at a pressure of 1 hPa as a function of the driving current and heat-sink temperature of the QCL (660 μm long QCL emitting at 3.1 THz) (b) Transmission signal as function of the driving current at temperature 41 K (horizontal line) (c) Transmission signal as function of the heat-sink temperature at 350 mA (vertical line)

4.6.3 Frequency calibration

Molecular absorption lines offer an obvious and readily available frequency standard in the THz range. For a well-known species, the individual lines of the spectral fingerprint can be identified according to a molecular catalog. The spectra are measured for that particular species and then compared to the data in the molecular databases. CH_3OH has been used as a probe species for most of this work, and the Jet Propulsion Laboratory (JPL) database [90] is used for frequency calibration. The spectrum shown in Figs. 4.10(a) and 4.10(b) has been obtained by normalizing the transmission signal shown in Figs. 4.9 (b) and 4.9(c). For the normalization, the data from Figs. 4.9(b) and 4.9(c) is divided by the raw data for an empty absorption cell. The frequencies and intensities for the observed lines were compared with values from the JPL database. After comparing the detected lines with the molecular catalog, a single-mode frequency coverage of 2.39 GHz was inferred.

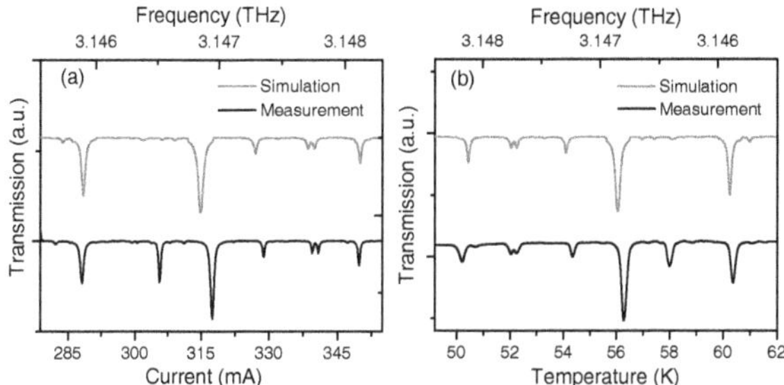

Figure 4.10: (a) Transmission signal due to current tuning (b) Transmission signal due to temperature tuning after the baseline correction of the signal shown in Fig. 4.9(b) and 4.9(c). The black line represents the measured spectra, while the red line is the result of the simulation.

Figures 4.10 shows the individual absorption lines of CH_3OH as a function of the QCL current and heat sink temperature. With such a frequency calibration, the tuning behavior can be precisely determined, and the QCL frequency ν is well described by a linear dependence on the driving current and the heat sink temperature. For current tuning, it was found that when the QCL driving current increased, the QCL frequency changed as

$$\nu = \nu_0 \pm a_I I \tag{4.1}$$

While for temperature tuning, the increasing temperature of the heat sink tuned the QCL frequency ν as

$$\nu = \nu_0 \pm a_T T \tag{4.2}$$

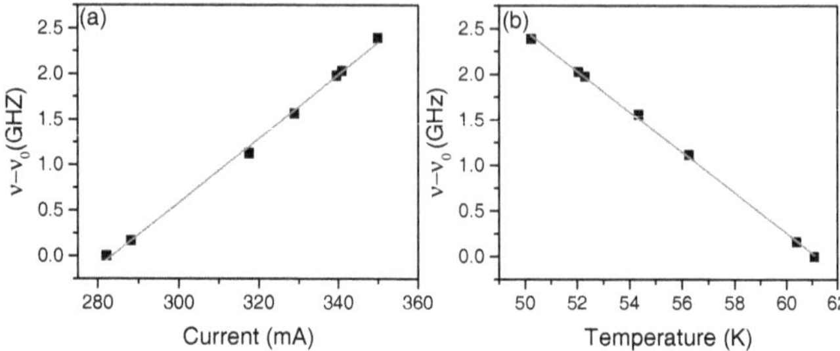

Figure 4.11: Frequency dependence due to (a) current tuning (b) temperature tuning. The black squares represent the measured absorption lines, while the red line is the linear fit, where $\nu_0 =$ 3145.7341 GHz.

The parameter ν_0 represents the unperturbed QCL frequency, a_I, and a_T are the proportionality constant for the QCL current and heat sink temperature, respectively. Figs. 4.11(a) and 4.11(b) shows the measured absorption lines fitted with the data from JPL. Black squares refer to the frequencies of the CH_3OH absorption lines, and the solid red line refers to the linear fit to the data according to Eqs. 4.1 and 4.2. From the fit, the frequency tuning coefficient calculated for the current is $d\nu/dI = +(35 \pm 0.75)$ MHz/mA and the temperature is $d\nu/dT = -(220 \pm 3.0)$ MHz/K for this device.

Chapter 5

Doppler free spectroscopy with a THz QCL

Nonlinear laser spectroscopy enables to overcome limitations in molecular spectroscopy due to Doppler broadening. The absorption of laser radiation by molecules changes the population density of the involved states. At larger intensities, this leads in turn to saturation and nonlinear absorption features. A prominent example is the observation of a Lamb dip in the center of an inhomogeneously broadened absorption line, which can be exploited for high-resolution spectroscopy and frequency stabilization. In order to perform nonlinear saturation spectroscopy, lasers with narrow-band emission are required. At THz frequencies, QCLs are suitable because they are powerful continuous-wave sources with a quantum-limited intrinsic linewidth of the order of 100 Hz [89]. This makes them ideal sources for spectroscopic applications requiring ultimate frequency resolution as in the fields of THz metrology and THz remote sensing of gaseous species. While in recent years, Doppler-free saturation spectroscopy based on QCLs has been successfully established in the mid-infrared spectral region [91, 92, 93], previous results in the THz range are solely based on upconversion of microwave sources [94, 95]. While optical saturation with a QCL was already reported in an earlier study for a transition of methanol at 2.5 THz, no Doppler-free spectra were obtained [96]. Due to the much narrower absorption lines as compared to the mid-infrared range, challenges for Doppler-free spectroscopy are the required pump power in combination with a sufficiently small linewidth of the laser source and a sensitive detection method. This chapter will discuss the evidence of saturation and Doppler-free spectroscopy with a THz QCL.

5.1 Theoretical background

Doppler-free (or Lamb dip) spectroscopy is based on the velocity-selective saturation of Doppler broadened molecular transitions. The mechanism of this type of spectroscopy can be understood simply with the help of a pump-probe configuration. When a laser beam passes through an absorption cell at a given frequency ν but not at the central resonant frequency ν_0 ($\nu \neq \nu_0$), the incident wave (probe beam) is absorbed by molecules with the velocity components $V_z = +(\nu - \nu_0 \pm \Gamma/2)/k$, the reflected beam is absorbed by other molecules with $V_z = -(\nu - \nu_0 \pm \Gamma/2)/k$, where k is the wave vector [69] and Γ the full width at half maximum (FWHM) of the transition in the absence of Doppler broadening, i.e., the sub-Doppler linewidth. However, if the incident laser radiation is at the central resonant frequency ($\nu = \nu_0$), both the pump and probe beams will interact with the same subset of molecules that have no component of velocity in the z-direction. The intensity per molecule absorbed is now twice as large, and the saturation accordingly higher. In such a configuration with counter-propagating waves, the pump intensity (I_{pump}) is kept at the line center (ν=ν_0) to saturate the molecular line, and the weak probe intensity (I_{probe}) is tuned across that line profile. The observable Lamb dip is well described by the rate equation approach described in, Letokhov and Chebotayev [97]. In the limit of $I_{probe} \ll I_{pump} \ll I_S$, where I_{probe} denotes the probe intensity, the absorption coefficient of the probe is given by

$$\alpha = \alpha_0 \Big(1 - \frac{S}{2} \frac{(\Gamma/2)^2}{(\nu - \nu_0)^2 + (\Gamma/2)^2}\Big). \tag{5.1}$$

Here, S denotes the saturation parameter. The saturation parameter is given by $S = I_{pump}/I_S$. α is the integrated-absorption coefficient for inhomogeneous broadening (Doppler broadening)

$$\alpha = \frac{\alpha_0}{\sqrt{1 + I_{pump}/I_S}}, \tag{5.2}$$

where α_0 the unsaturated absorption coefficient, I_{pump} the pump intensity, and I_S the saturation intensity. For larger pump intensity (I_{pump}) the parameters S and Γ in Eq. 5.1 have to be replaced by

$$S \rightarrow \tilde{S} = \frac{2S}{1 + S + \sqrt{(1+S)}}. \tag{5.3}$$

$$\Gamma \rightarrow \tilde{\Gamma} = \frac{\Gamma}{2}\Big(1 + \sqrt{1+S}\Big). \tag{5.4}$$

Hence, the spectral width of the Lamb dip depends on the saturation parameter, which is known as power broadening. If the linewidth of the laser is not negligible with respect to the Lamb dip, the absorption has to be convoluted with the normalized line profile I(ν) of the laser ($\int_0^\infty I(\nu)d\nu = 1$):

$$\alpha_{eff}(\nu) = (\alpha * I)(\nu) \tag{5.5}$$

The linewidth of the Lamb dip will become minimum in case of a negligible laser linewidth and a small pump power (negligible power broadening). Note that both, the width and the depth of the Lamb dip, are affected by Γ through the convolution according to Eq. 5.5, since the spectral area of the Lamb dip, which is proportional to $\tilde{\Gamma}, \tilde{S}$ is conserved.

5.2 Experimental setup for QCL characterization and saturated absorption spectroscopy

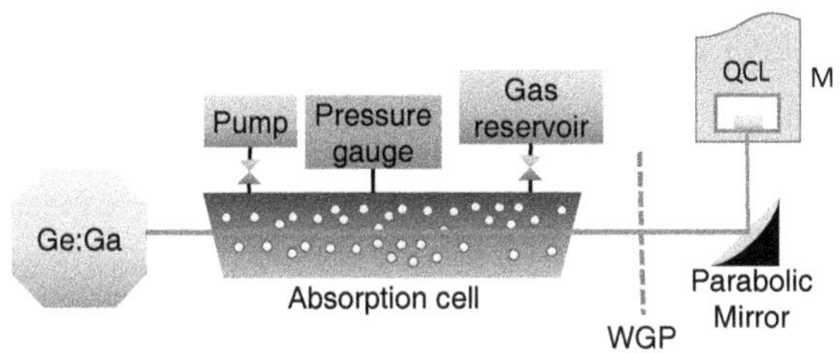

Figure 5.1: Schematics setup for linear absorption spectroscopy and saturation measurements. WGP: Wiregrid polarizer; M: Mechanical cryocooler.

The QCL used in this experiment is based on a $GaAs/Al_{0.18}Ga_{0.82}As$ active region heterostructure and a single plasmon waveguide [53]. The laser emits at 3.3 THz up to 1 mW of cw output power. Initially, the QCL emission is characterized by Fourier-transform spectroscopy to verify single-mode operation and to determine the emission frequency with a precision of about 5 GHz (setup shown in Fig. 4.7). In a second step, direct absorption spectroscopy is performed for a more precise frequency calibration based on methanol (CH_3OH) with its well-documented fingerprint spectrum as a reference [90]. The experiment is designed to characterize the QCL and to measure the appropriate parameters for saturation spectroscopy. The corresponding setup is illustrated in Fig. 5.1. The QCL is operated in a continuous liquid-He flow cryostat and stabilized at a temperature of 37 K. The beam was collected using an off-axis parabolic mirror, and the transmitted radiation is measured by a Ge:Ga detector through an absorption cell.

A scanning rate of 1–2 A/s has been applied to the QCL to reduce low-frequency flicker noise and to maintain sufficient frequency resolution for Doppler-free measurements. The current is tuned from below the threshold over the entire dynamic range of single-mode QCL operation, resulting in an acquisition time of 100–200 ms/scan. A wire grid polarizer is used in front of the absorption cell to control the power of the incident beam. In the final step, The QCL beam profile is measured by a microbolometer camera, and the output power is measured by Thomas Keating (TK) power meter.

5.3 Results

The absorption spectrum is shown in Fig. 5.2(a) is obtained by sweeping the driving current of the QCL at a temperature of 37 K. CH_3OH is used as a probe gas, and the spectrum is measured at a pressure of 1 hPa. By comparing the data to the JPL catalog, a single-mode frequency range of 3.310–3.313 THz is deduced with tuning parameters of -170 MHz/K and $+8$ MHz/mA. For saturation spectroscopy, as mentioned in Subsect. 3.6.4, typically a much lower pressure is required, and the absorption line to be examined should be observable at this pressure. It is observed that CH_3OH transitions at this frequency range were not strong enough to perform Lamb dip experiments. The molecular species is therefore changed to deuterate water (HDO) as HDO is found to have a strong transition in this frequency region. In the laboratory, H_2O was deliberately mixed with D_2O to produce the HDO, and the chemical equation is:

$$H_2O + D_2O \rightleftharpoons 2\,HDO \tag{5.6}$$

The measurement result for the HDO is shown in Fig. 5.2(b). The particularly strong rotational strong line is identified as the $J = 10_{4,7} \leftarrow 10_{3,8}$ transition of single-deuterated water (HDO) at 3.311835 THz [90]. The FWHM value of the transition line is around 9.4 MHz at a pressure of $\sim$1 Pa.

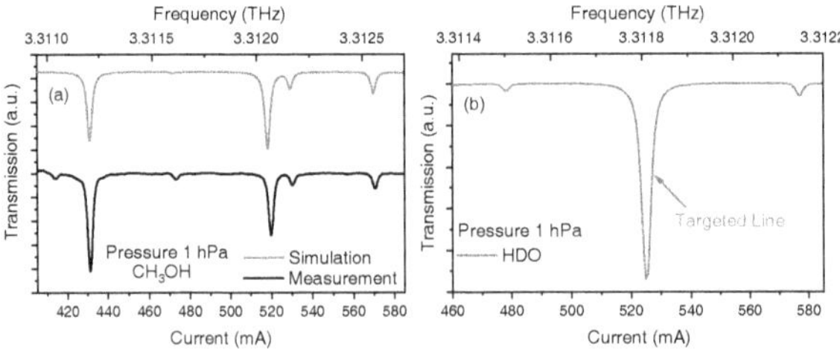

Figure 5.2: (a) Spectral measurement of CH_3OH and compared with a JPL simulation (b) Absorption spectra for different isotopic compositions.

Initially, the experimental saturation parameter S and the unsaturated absorption coefficient α_0 has been calculated from a power-dependent measurement while attenuating the pump power in front of the gas cell with a rotating wire-grid polarizer. Fig. 5.3(a) shows the absorption profile at a pressure of 1 Pa measured at a QCL's heat-sink temperature of 37 K. As can be seen, the absorption decreases as the QCL intensity increases due to the presence of saturation. In the absence of saturation, at a higher pressure, the absorption profile does not change with intensity. The evidence of the saturation effect is more clearly demonstrated by plotting the absorption coefficient as a function of the pump intensity for different values of the total pressure (Fig. 5.3(b)). The dotted lines refer to the absorption coefficient as a function of the pump intensity, and the solid lines refer to fit Eq. 5.2. As expected, the absorption becomes less saturated toward higher pressures. The data plotted in Fig. 5.3(b) also allow quantitative estimation of the saturation intensity for the investigated transition. For the lowest pressure of 0.25 Pa, which is also used for Doppler-free experiments, the values of $\alpha_0 = (7.4 \pm 0.4) \times 10^{-4}$ cm^{-1} and $I_S = (0.23 \pm 0.05) I_0$ are determined from Eq. 5.2, where I_0 denotes the unattenuated pump intensity. By comparing the unsaturated absorption coefficient to the molecular database, the partial pressure of (0.058±0.030) Pa for HDO is determined. This value differs from the total pressure due to the coexistence of non- and double-deuterated water molecules.

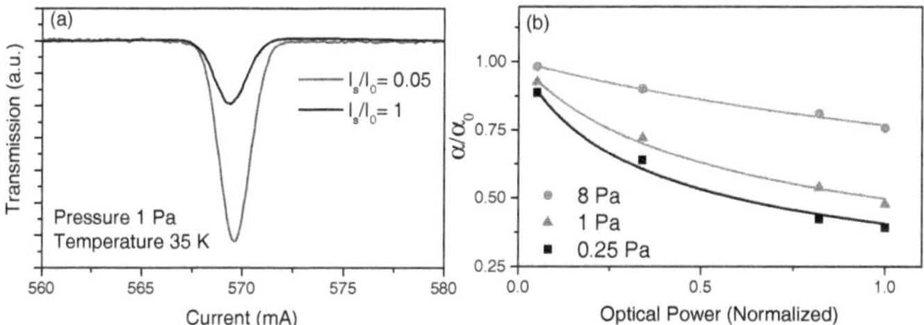

Figure 5.3: (a) Absorption spectrum of HDO mixer acquired at a pressure of 1 Pa and with different QCL intensities (b) Normalized absorption coefficient α/α_0 as a function of the normalized pump intensity, I_{pump}/I_0 for different values of the total pressure. The solid lines refer to a fit to Eq. 5.2 with the following parameters: α_0=7.4×10^{-4} cm^{-1} and I_s/I_0=0.23 for 0.25 Pa; α_0 = 3.9×10^{-3} cm^{-1} and I_s/I_0 = 0.53 for 1 Pa; α_0 = 2.2×10^{-2} cm^{-1} and I_s/I_0 = 2.1 for 8 Pa.

To determine the power spectral density for this particular measurement, the beam profile and the corresponding power is measured after the gas cell. The acquired beam profile with a microbolometer camera is shown in Fig. 5.4. The blue and black dotted Gaussian curves in Fig. 5.4 (b) and 5.4 (c) display the horizontal and vertical cross-section, respectively of the beam profile, corresponding to the profile shown in Fig. 5.4 (a). The dotted lines refer to the measurement data, and the solid lines are the result of a Gaussian fit. The intensity distribution is well described by a Gaussian profile, and the measured beam waist for that particular optical configuration is around w_x = 2.66 mm and w_y = 2.8 mm. From total power measurements with the THz power meter (TK-THz-PM07), a maximum pump power of 0.76 mW was measured after the gas cell, corresponding to a peak pump power density of 65 μW/mm^2.

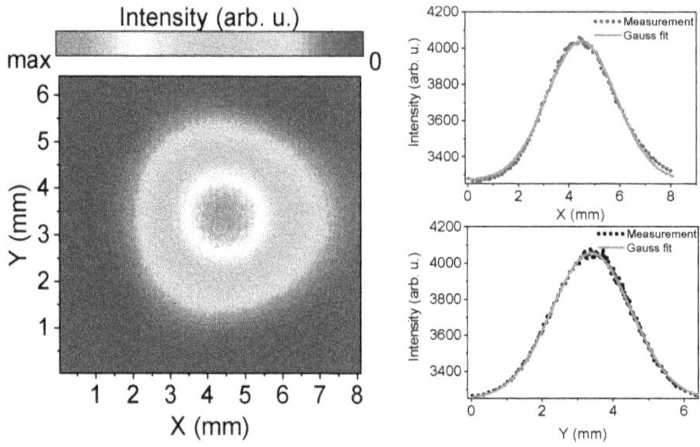

Figure 5.4: Intensity profile of the QCL at the beam waist position. (b) Horizontal cross section, and (c) Vertical cross section of the beam profile. The dotted lines refer to the measurement data, and the solid lines are the result of a Gaussian fit.

5.4 Experimental setup for Lamb-dip spectroscopy

After these initial experiments, the setup is modified for the Lamb dip measurements. Lamb dip experiments require two counter-propagating beams of the same laser. The schematic configuration and photograph of the lab setup of the experiment are shown in Figs. 5.5 (a) and Fig. 5.5 (b). A collinear pump-probe setup is generated by a plane mirror after the gas cell. One of the main limiting factors for this measurement is the optical feedback from the counter-propagating probe beam, which induces QCL frequency instabilities. A wire grid polarizer (WGP) is used in combination with a quarter-wave ($\lambda/4$) plate made from x-cut quartz to deflect the probe beam to the detector and to regulate the impact of external optical feedback to a tolerable amount. The absorption cell is sealed in such a way that data can be obtained at very low pressures. Due to detection noise, an averaging over several sweeps is performed for a reliable quantitative analysis of the data. The injection current of the laser is tuned by a low-noise current driver (Wavelength Electronics, QCL 1000 OEM).

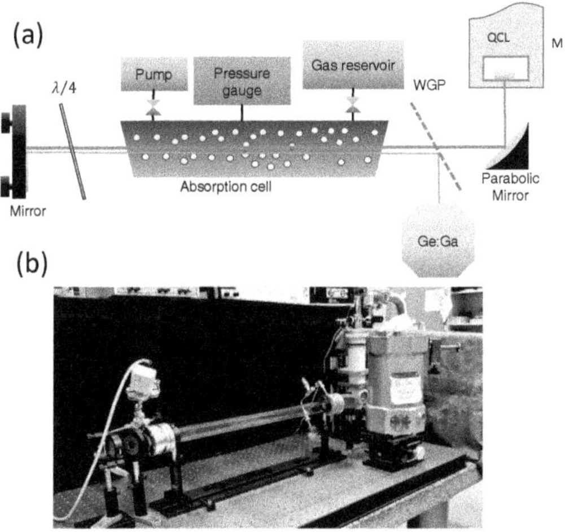

Figure 5.5: (a) Schematic setup; (b) photograph of the lab setup for Lamb dip spectroscopy

5.5 Results

The experiment was carried out using the strongest observable transitions of deuterated water (HDO) ($J = 10_{4,7} \leftarrow 10_{3,8}$) in the spectral range covered by the QCL. The Lamb dip was monitored for different gas pressures (0.25–1 Pa) inside the absorption cell. As the gas pressure within the cell decreases, the size of the Lamb dip increases. The results at a pressure of 0.25 Pa are shown in Figs. 5.6(a)–(c) at various pump intensities. The blue lines refer to the experimental data, and the red lines display results of the simulations according to Eqs. 5.1–5.5 assuming a Gaussian lineshape for the QCL. Figures 5.6(d)–(e) depict the observed Lamb dip after simply subtracting a Gaussian from the experimental data. The observed Lamb dip becomes deeper with increasing pump power. This is an expected result since the increased optical power within the cell increases the saturation parameter, S. The value of Γ, as well as the linewidth of the laser, can be estimated by analyzing the data since both affect the shape of the Lamb dip according to Eqs. 5.1–5.5. For a saturation parameter of $S = 4.3$ as obtained from Fig. 5.3(b) for pressure at 0.25 Pa, a sub-Doppler linewidth of $\Gamma = 170 \pm 30$ kHz and an FWHM of $\gamma_{QCL} = (750 \pm 150)$ kHz was calculated from the data fit. The effective linewidth of the QCL is hence much larger than the intrinsic Shawlow-Townes limit, which can be below 100 Hz for THz QCLs [89].

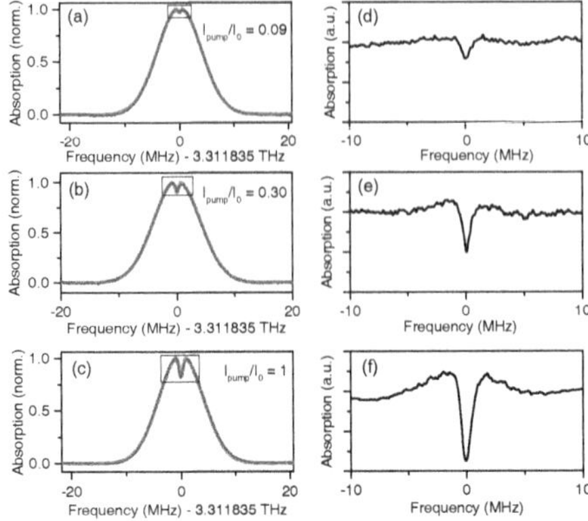

Figure 5.6: (a)–(c) Normalized absorption signal for different intensities of the pump beam for a total pressure of 0.25 Pa. The blue lines refer to the experimental data (averaged over 19 scans with a current scan rate of 1 A/s at 30 K), while the red lines lines show the results of simulations. The simulation parameters are: $S = 4.3$, $\Gamma = 170$ kHz, $\gamma_{QCL} = 700$ kHz [$\gamma_{QCL} = 800$ kHz in (c)]. (d)–(e) Observed Lamb dip after subtracting a Gaussian fit from the experimental data of (a)–(c).

The QCL emission linewidth depends on the measurement timescales. The large linewidth of the QCL is mainly due to the limited stability of the driving current and the temperature. Here is a brief discussion on the contribution of current and temperature noise to the QCL linewidth of the experiment:

- The impact of temperature fluctuations can be estimated from the temperature tuning coefficient of 170 MHz/K. The linewidth of 750 kHz corresponds to the temperature stability of ± 2.2 mK, which is compatible with the stability of the temperature control loop used in this experiment. Due to the heat capacity of the components, temperature fluctuations can be reduced on shorter timescales. A temperature-related frequency change of 100–400 kHz for 20–50 ms acquisition time is calculated based on the temperature tuning coefficient of the QCL.

- The frequency resolution is limited by the sweeping rate and the speed of acquisition. For a sweep rate of 1 A/s and a sampling rate of 50 kHz, the step-size is 22 μA, which corresponds to about 100 kHz. This is well above the intrinsic noise level of the current driver (Wavelength Electronics, QCL1000), which is specified as 1 μA rms for a bandwidth of 100 kHz.

The experimental value of the Lamb dip has a width of 0.7 ± 0.2 MHz, but the calculated sub-Doppler linewidth is 170 kHz. The discrepancy between the calculated value of Γ and the experimentally determined value of Γ can be explained by three factors which have an effect on the sub-Doppler linewidth: the transit time ($\Delta\nu_{transit}$), pressure ($\Delta\nu_p$), and power broadening ($\Delta\nu_s$). The total width of the Lamb dip can be estimated from $\delta\nu_{LD} = \sqrt{\Delta\nu_{transit}^2 + \Delta\nu_p^2 + \Delta\nu_s^2}$. The transit time broadening $\Delta\nu_{transit}$ = 0.4 v/w ≈ 50 kHz, where, beam waist, w = 2.8 mm and mean velocity of molecule, v = 360 m/s at 300 K [69]. A pressure broadening of 40 kHz is determined according to the pressure-broadening coefficient of 16 MHz/hPa. Power broadening, which is around 697 kHz, contributes the most to the Lamb dip spectrum. The primary cause for power broadening is the power of the pump beam required to obtain a sufficient signal to noise ratio.

5.6 Doppler-free spectroscopy inside a mechanical cryocooler

Mechanical cryocoolers are an innovative technology for precision space instruments, and most of the application scenarios rely on mechanical cryocoolers instead of flow cryostats. So finally, the Doppler-free measurement was carried out inside a mechanical cryocooler. The same optical configuration is used here, as shown in Fig. 5.5, but a mechanical cryocooler replaces the He-flow cryostat. One of the disadvantages of using a mechanical cryocooler is that it produces large amounts of vibration. In this measurement, the vibrations are dominated by the 45-Hz piston cycle of the employed Stirling cooler (Ricor, model K535), which has been identified as a source of frequency fluctuation. A Lamb dip signal is still observable under such conditions if the measurement is synchronized to the fundamental frequency of the cooler vibration. Here, a synchronization technique was developed to eliminate the influence of the cooler vibration and to observe the Lamb dip. In order to synchronize the measurement device, the fundamental frequency was obtained from the cooler drive electronics. The frequency signal was then fed to the programmable function input (PFI) of the compact DAQ chassis using a step-down transformer. By synchronizing the measurement with the fundamental frequency of the piston cycle, the signal-to-noise ratio improved, and the Lamb dip was observed quite profoundly. The remaining fluctuations at higher-order vibration frequencies can be tolerated if the acquisition time for the spectral range of interest is made sufficiently short. Vibration-induced flicker noise does then mainly appear as a scan-to-scan frequency offset in the spectra, which can be compensated by a post-acquisition jitter correction routine. One of the limiting factors is the sampling rate of the acquisition and control hardware (50 kHz), which restricts the maximum current scan rate to about 2 A/s in order to maintain a sufficiently small frequency step for Lamb dip spectroscopy.

The optical setup, as shown in Fig. 5.5 suppresses optical feedback already to a large extent. However, residual feedback might be caused by unintended reflections as well as by the limited efficiency of the quarter-wave plate and the wire grid polarizer. It is seen that even rather strong optical feedback can be tolerated in this optical setup. This is due to a peculiarity of the THz range, where the free spectral range of the external cavity can be large compared to the Doppler width of a molecular absorption line. The level of optical feedback can be adjusted by rotating the quarter-wave plate. One indication of optical feedback is then found in the operating voltage of the QCL, which contains a feedback-related self-mixing component. Figure 5.7(a) depicts the first derivative of the QCL voltage in the presence of optical feedback. A periodic self-mixing component is found, where the periodicity reflects the 175-MHz free spectral range of the optical cavity formed by the QCL and the mirror after the gas cell. The corresponding absorption signal after third-order polynomial baseline subtraction is depicted in Fig. 5.7(b). The Lamb dip is clearly observed within the absorption peak. Two additional features (at 555 and 578 mA) are observed in the absorption signal, which is correlated to the self-mixing voltage spikes in Fig. 5.7(a) and caused by the external-cavity resonances.

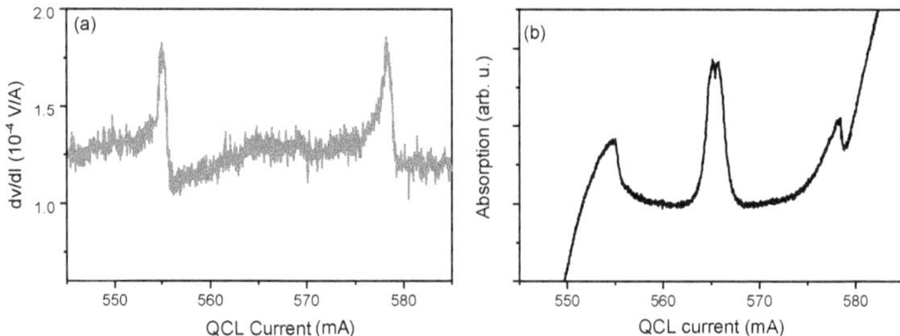

Figure 5.7: Lamb dip spectroscopy in the presence of optical feedback. The QCL was operated in a Stirling cooler at 35 K. (a) First derivative of the measured QCL voltage (single scan, 2 A/s) and (b) measured detector signal after baseline subtraction as a function of the driving current.

Figure 5.8(a) shows the HDO absorption line at a pressure of 1.6 Pa with a pronounced Lamb dip in the center of the transition. The absorption profile is best suited to the Gaussian profile, and the measured Lamb dip has a width of 0.6–0.8 MHz. The frequency noise between subsequent scans is observed due to temperature drift and cooler vibration. Also, the frequency instability during the scan induced the mechanical jitter in the cavity, and the low-frequency noise in the current controller made it challenging to detect the Lamb dip. In order to reduce this noise in the absorption spectra, an average of multiple scans is performed. A least-square algorithm has been performed in order to correct the offset, which minimizes the quantity $\sum_i^N \left(S_i - S'_{i+j}\right)^2$ as a function of the offset parameter j. Here, S_i and S'_i denote subsequent sweep spectra. The Fig. 5.8(b) is shown for averaging of 11 subsequent sweeps. While the Lamb dip is clearly seen for the offset correction method, it becomes completely obscured without offset correction.

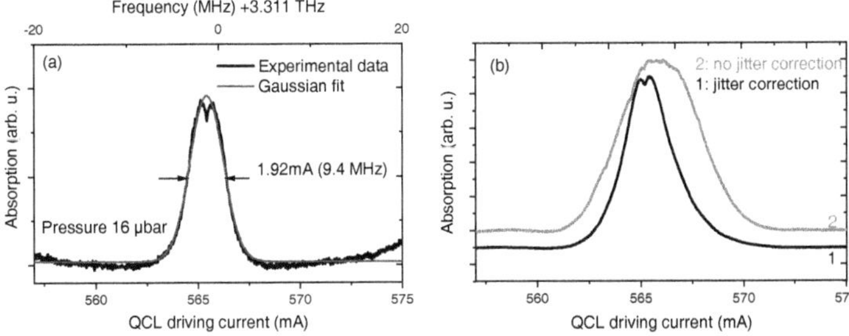

Figure 5.8: (a) Absorption line of HDO exhibiting a pronounced Lamb dip (b) Absorption signal as a function of the driving current 1: averaging over 11 sweeps with offset correction; 2: averaging without offset correction.

5.7 Conclusion

Sub-Doppler spectroscopy based on a THz QCL and a free-space gas cell in a collinear pump-probe configuration is demonstrated. The results clearly show that a sub-MHz laser linewidth is achievable for measurements with a free-running QCL even for integration times of several seconds. In the case of operation in a vibration-free flow cryostat, the QCL linewidth becomes limited by the achievable temperature and current stability. It has also been found that a certain level of stationary optical feedback is tolerable for Doppler-free measurements as long as the free spectral range for the optical cavity is large compared to the width of the molecular absorption line. More specifically, these experiments have demonstrated the high power and narrow linewidth of the THz QCLs. Overcoming the Doppler-limited resolution of the QCL-based THz spectrometer could improve its precision by even two orders of magnitude. Finally, It might enable for future Lamb dip stabilization based on devices that are phase-locked to an electronic or optical frequency reference and reveal a wide variety of sensing applications.

Chapter 6

Molecular spectroscopy by light-induced frequency tuning of THz QCLs

THz spectroscopy is a powerful tool for optical metrology, astronomy, and atmospheric research because of their substantial number of molecular rotational transitions and atomic fine-structure lines in their fingerprint region [1–10 THz]. Spectroscopy of such transitions relies on broad frequency-tunable sources with a linewidth in the MHz or sub-MHz range. Commercially available electronic sources such as multiplied microwave oscillators typically provide a frequency coverage of about ten percent of the center frequency. However, the output power of such available sources drops to a few tens of microwatts as the frequency increases above 1 THz. QCLs are therefore of particular relevance for spectroscopy above 2 THz because they can deliver milliwatts of output power with intrinsic linewidth limit in the sub-kHz range [51, 65, 57, 89]. However, so far, their frequency tuning is somewhat limited. The frequency tuning, which can be realized by varying the driving current and/or the heat sink temperature is usually well below 10 GHz. Since many of the spectroscopic methods based on THz QCLs will immediately benefit from a larger tuning range [82, 84, 98, 99, 100], there is a strong motivation on developing methods which enhance the frequency tuning capability of these lasers.

Several techniques have been demonstrated during recent years. Among them are external optical cavities with reported continuous tuning ranges within 12–50 GHz [101, 102], micro-optomechanical cavities with 67–240 GHz [103, 104], electronic frequency tuning in multi-terminal and multi-section QCLs with 2–19 GHz [105, 106, 107], and tuning by gas condensation with 25 GHz [108] as well as by cavity pulling effects in microdisk resonators [109]. However, none of these approaches have been established so far for high-resolution spectroscopy, which may originate from the often involved complicated setups and device fabrication procedures or from non-continuous or non-reproducible frequency tuning characteristics. Near-infrared (NIR) laser excitation was exploited by several groups to manipulate the output power and frequency of mid-infrared QCLs [110, 111, 112]. Recently, a significant frequency tuning of already several GHz for cw operation was obtained for THz QCLs by unfocused NIR illumination [113]. This chapter demonstrates the feasibility of a wide-band frequency tuning of THz QCLs using a spatially controlled NIR illumination. A physical model for the observed frequency tuning characteristics is described in the following section.

6.1 Theoretical background

A one-dimensional Fabry-Pérot cavity consisting of GaAs is considered to represent the physical model of the light-induced frequency tuning (LIFT) mechanism. By illuminating a semiconductor with light above the bandgap, electron-hole pairs are generated. Immediately after generation, these pairs form a plasma state of unbound electrons and holes, which subsequently decays via various channels. As operating temperatures are below 100 K, the diffusion of the carriers is neglected. Further, the surface and the Auger recombination are neglected, and the electron density n_e is assumed to be equal to the hole density n_h. In the steady-state, the plasma density $n_{eh} = n_e = n_h$ is then described by [114]

$$\frac{n_{eh}}{\tau_{nr}} + B_r n_{eh}^2 = G, \tag{6.1}$$

where G, τ_{nr}, and B_r denote the generation rate, the non-radiative lifetime and the bimolecular recombination parameter, respectively. At low temperatures, non-radiative recombination is strongly reduced [115], while bimolecular recombination becomes enhanced [116]. The concentration of the electron-hole plasma becomes approximately

$$n_{eh} = \sqrt{\frac{G}{B_r}}. \tag{6.2}$$

Due to absorption, G decreases exponentially in the semiconductor with

$$G(z) = G_0 e^{-\alpha z}. \tag{6.3}$$

Here, G_0 denotes the generation rate at the surface, α the absorption coefficient at the particular excitation wavelength, and z the distance from the facet along the waveguide. Reabsorption of the undirected luminescence, which might cause a small reduction of the effective absorption coefficient, is neglected for the sake of simplicity. The generation rate per unit volume is related to the incident photon flux S_0 (number of photons per time and area) by $G_0 = \alpha S_0$. The presence of the electron-hole plasma changes the refractive index of the semiconductor. The local dielectric constant is given by

$$\epsilon(z) = \epsilon_s(1 - \frac{\omega_p^2}{\omega^2}), \tag{6.4}$$

where $\epsilon_s = n_s^2$ denotes the dielectric constant of the GaAs substrate without excitation, $\omega = 2\pi\nu$ the angular frequency of the QCL, and ω_p the plasma frequency which is given by

$$\omega_p^2 = \frac{n_{eh}(z)e^2}{\epsilon_0 \epsilon_s m^*}. \tag{6.5}$$

Here, e denotes the electron charge, ϵ_0 the vacuum permittivity, and m^* the effective electron mass. For simplicity, ϵ_s and n_s are assumed as real-valued. Equation 6.4 is plotted for different photon fluxes in Fig. 6.1, which demonstrates the behavior of the dielectric constant in the substrate underneath the facet. For small fluxes, only the real part of the refractive index in the vicinity of the facet ($z = 0$) is affected. Above a certain photon flux, a transition from a dielectric to a metallic behavior takes place at the position where the plasma frequency ω_p equals the QCL frequency. With increasing photon flux, this transition moves deeper into the cavity. The resulting shift of the cavity resonance can be numerically calculated with the transfer matrix method by considering the cavity as a stack of stratified layers in z direction [117].

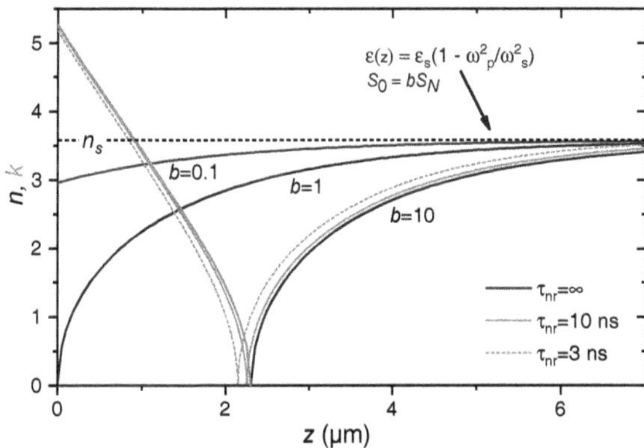

Figure 6.1: Calculated change of the real part of the refractive index close to the facet of the excited GaAs substrate for normalized incident photon fluxes $S_0 = bS_N$, where S_N is the photon flux for which $\epsilon = 0$ at the surface ($z = 0$). For $b = 10$, the calculated change of the imaginary part of the refractive index is also shown, as well as further results for different values of τ_{nr}.

In order to calculate the cavity resonance shift, a resonator is considered to have a length of L and is divided into two sections A and B, where B is the section that is optically excited. An analytical approximation for the shift of the transmission maxima will be derived, which will eventually yield the tuning equation.

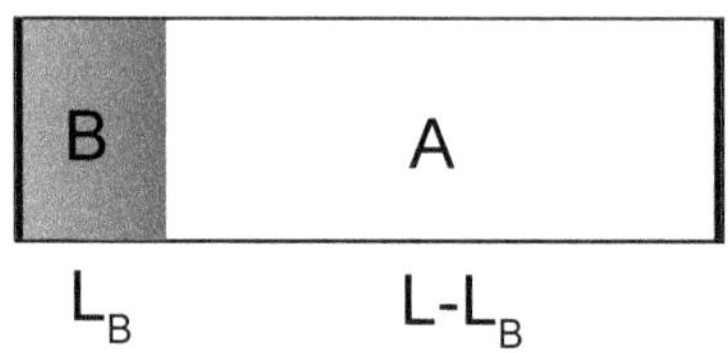

Figure 6.2: Resonator with a length of L

Since the plasma excitation takes place on a length scale much smaller than the wavelength of the QCL and since the change of the dielectric constant can be very large, the problem cannot be covered in the framework of a scalar Eikonal equation. The following derivation starts with the rather general transfer-matrix formalism for stratified media, as it is routinely used to calculate, for instance, the transmission and reflection spectra of layered structures [117]. Within this framework, the relation of the electric and magnetic field strengths E_1 and H_1 on the left-hand side of the resonator with E_0 and H_0 on the right-hand side is given by two coupled equations, which are linear with respect to the field components. For a single dielectric layer under normal incidence, the matrix notation of these equations reads

$$\begin{pmatrix} E_1 \\ H_1 \end{pmatrix} = \begin{pmatrix} \cos(k_l l) & iZ_l \sin(k_l l) \\ \frac{i}{Z_l}\sin(k_l l) & \cos(k_l l) \end{pmatrix} \begin{pmatrix} E_0 \\ H_0 \end{pmatrix}. \tag{6.6}$$

Here, $Z_l = \sqrt{\mu_0/\epsilon_l \epsilon_0}$ denotes the wave impedance of the layer, l the thickness, and $k_l = n_l k_0$ the wavevector. In order to avoid imaginary matrix elements, a base transformation is performed $(E_i, H_i) \rightarrow (\tilde{E_i}, \tilde{H_i})$ with $\tilde{E_i} = -i\frac{\sqrt{\epsilon_0}}{\sqrt{\mu_0} k_0} E_i$ and $\tilde{H_i} = H_i$. The transfer matrix becomes then

$$M_l = \begin{pmatrix} \cos(k_l l) & \frac{1}{k_l}\sin(k_l l) \\ -k_l \sin(k_l l) & \cos(k_l l) \end{pmatrix}. \tag{6.7}$$

The transfer matrix M of the composed cavity is obtained as the product of the transfer matrices $\mathcal{A}$ and $\mathcal{B}$ referring to sections A and B

$$M = \mathcal{B} \cdot \mathcal{A} = \begin{pmatrix} M_{11} & M_{12} \\ M_{21} & M_{22} \end{pmatrix}. \tag{6.8}$$

The transmission coefficient for the cavity is given by

$$t = 2ik_0 e^{-ik_0 L} \frac{1}{-M_{21} + k_0^2 M_{12} + ik_0(M_{11} + M_{22})}. \tag{6.9}$$

To determine M, the transfer matrices $\mathcal{A}$ and $\mathcal{B}$ must be found first. Subsequently, the complex roots of the denominator must be identified in Eq. 6.9 with respect to the cavity wavevector, since these represent the lasing condition. If the cavity is divided into a large number of stratified layers, the frequency shift can be calculated numerically using the transfer matrix method by finding the complex roots of t^{-1} or by simply determining the shift of the transmission resonance as the function of the excitation density (neglecting the influence of gain). Both approaches are possible without further approximations, but cannot provide a tuning equation. For deriving an analytic expression, we are interested in a simple form for the final transfer matrix M. Consequently, all small contributions of higher-order will be neglected in the following.

For the determination of the transfer matrix $\mathcal{B}$, the length of section B is assumed to be smaller than the THz wavelength $L_B \ll \lambda$ and that the optical excitation is completely absorbed in this section. In Eq. 6.7, the cosine is replaced by 1, and the sine is replaced by its argument. To take into account the change of the dielectric constant by optical excitation, section B is divided into small subsections with a thickness of Δz. The matrix $\mathcal{B}$ is obtained as the product of all sub-matrices by performing the transition to infinitesimal small layer thicknesses dz.

$$\mathcal{B} = \prod_{\Delta z \to 0} \begin{pmatrix} 1 & \Delta z \\ -k_B^2 \Delta z & 1 \end{pmatrix} = \begin{pmatrix} 1 & L_B \\ -\int k_B^2(z) dz & 1 \end{pmatrix}. \tag{6.10}$$

With Eq. 6.4, it can be written

$$\mathcal{B} = \begin{pmatrix} 1 & L_B \\ -k_s^2 L_B + k_s^2 \int \frac{\omega_p^2(z)}{\omega^2} dz & 1 \end{pmatrix}. \tag{6.11}$$

with $k_s^2 = \epsilon_s k_0^2$. Since it will simplify the further derivation significantly without affecting the final result, first a formal transition $L_B \to 0$ is carried out

$$\mathcal{B}^{L_B \to 0} = \begin{pmatrix} 1 & 0 \\ k_s^2 \int \frac{\omega_p^2(z)}{\omega^2} dz & 1 \end{pmatrix}. \tag{6.12}$$

The higher-order terms in L_B are neglected in the final approximation.

In order to find the transfer matrix for section $\mathcal{A}$, the system is considered to be close to resonance. So, one can write $k_sL = 2\pi N + \delta k_s L$ with N denoting an integer. Since optical gain is required in order to fulfill the lasing condition, δk_s has to be complex. This determines $\delta k_s = \Delta k_s - ig/2$ with Δk_s denoting the real component of the wave vector shift and g the threshold gain [114]. Assuming $|\delta k_s L| \ll 1$, and the cosine is approximated by 1 and the sine by its argument which results in

$$\mathcal{A} = \begin{pmatrix} 1 & \frac{\delta k_s}{k_s} L \\ -\delta k_s k_s L & 1 \end{pmatrix}. \tag{6.13}$$

With Eqs. 6.12 and 6.13, the simplified transfer matrix of the total cavity becomes eventually

$$M = \begin{pmatrix} 1 & \frac{\delta k_s}{k_s} L \\ \left[k_s^2 \int \frac{\omega_p^2}{\omega^2} dz - \delta k_s k_s L\right] & \left[1 + \delta k_s k_s L \int \frac{\omega_p^2}{\omega^2} dz\right] \end{pmatrix}. \tag{6.14}$$

It is now necessary to determine the complex roots of the denominator in Eq. 6.9

$$0 = M_{21} + k_0^2 M_{12} + ik_0(M_{11} + M_{22}). \tag{6.15}$$

Inserting the matrix elements and dividing by k_s results in

$$0 = \delta k_s L\left(1 + \frac{k_0^2}{k_s^2}\right) - k_s \int \frac{\omega_p^2}{\omega^2} dz + 2i\frac{k_0}{k_s} + i\delta k_s k_0 L \int \frac{\omega_p^2}{\omega^2} dz. \tag{6.16}$$

In order to fulfill the equation, real and imaginary part on the right side have both to become zero. While the former will yield the frequency shift, the latter will yield an expression for the threshold gain. Since δk_s is complex, both equations are coupled, i.e. there is in principle a gain contribution to the frequency shift and vice versa. However, these contributions are small and we neglect it. Further higher-order coupling terms originate from the imaginary component in the relation $k_s = (n_s - i\frac{g}{2k_0})$ with $n_s = \epsilon_s$ which has to be used in the presence of gain. The corresponding higher-order coupling terms are neglected by the approximation $k_s \approx n_s k_0$. Eventually, the real part of Eq. 6.16

$$\frac{\Delta k_0}{k_0} = \frac{\Delta \nu}{\nu} \approx \frac{\varepsilon_s}{\varepsilon_s + 1} \frac{1}{L} \int \frac{\omega_p^2}{\omega^2} dz. \tag{6.17}$$

Since $\varepsilon_s \gg 1$ and for the sake of simplicity, the term $\varepsilon_s/(\varepsilon_s + 1)$ has been approximated by unity. Carrying out the integral by using

$$\omega_p^2 = \frac{n_{eh}(z) e^2}{\varepsilon_0 \varepsilon_s m^*} = \frac{e^2}{\varepsilon_0 \varepsilon_s m^*} \sqrt{\frac{G}{B_r}} = \frac{e^2}{\varepsilon_0 \varepsilon_s m^*} \sqrt{\frac{\alpha S_0}{B_r}} \sqrt{e^{-\alpha z}}, \tag{6.18}$$

eventually obtained

$$\Delta \nu \approx \frac{e^2}{2\pi^2 \varepsilon_0 \varepsilon_s m^*} \frac{1}{L \nu} \sqrt{\frac{S_0}{B_r \alpha}}. \tag{6.19}$$

where non-radiative recombination is neglected. Here, ν denotes the unperturbed QCL frequency and L the physical cavity length. The frequency shift according to Eq. 6.19 is illustrated in Fig. 6.3(b) as a red line. It resembles the experimentally observed square root dependence and is in reasonable agreement with the corresponding exact numerical simulations even for rather large excitation intensities. Therefore, Eq. 6.19 can be used as a general estimate for the tuning range of a LIFT configuration. The same figure depicts the numerical results for different values of τ_{nr}. Since non-radiative recombination causes a linear component in the tuning characteristics, which is experimentally not observed, it is assumed that the non-radiative contribution is small in this configuration (with $\tau_{nr} \leq$ 3 ns, where the particular value of τ_{nr} might vary for different GaAs substrates). If Eq. 6.19 is written in terms of the excitation power $P = \hbar\omega_{NIR} S_0 A$,

$$\Delta \nu \approx \frac{e^2}{2\pi^2 \varepsilon_0 \varepsilon_s m^* L \nu \sqrt{\hbar\omega_{\mathrm{NIR}} A B_r \alpha}} \sqrt{P} = a_L \sqrt{P}. \tag{6.20}$$

Here, $\hbar\omega_{NIR}$ represents the photon energy and A the excitation area. Eventually, the QCL frequency shift can be written as follows

$$\nu(P) = \nu_0 + a_L \sqrt{P}. \tag{6.21}$$

The parameter ν_0 represents the unperturbed QCL frequency.

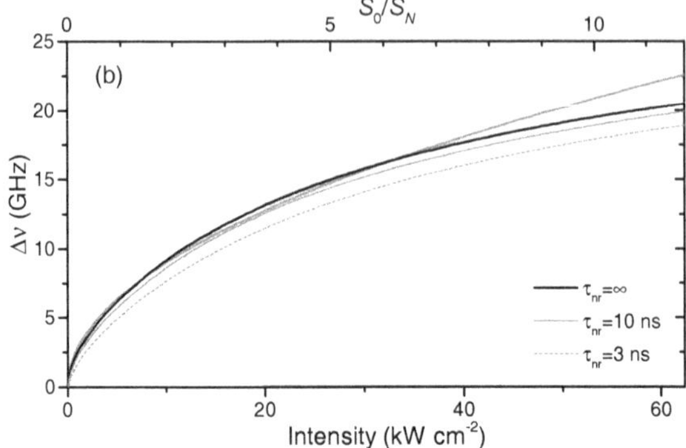

Figure 6.3: Calculated shift of the resonance frequency as a function of intensity ($\hbar\omega_{NIR}S_0$) for a one-dimensional cavity. Numerical results are depicted for different values of τ_{nr}. The excitation wavelength is 810 nm, L = 1 mm, ν = 3.3 THz, $\alpha = 10^4$ cm^{-1}, ϵ_s = 12.8, and $B_r = 1.8 \times 10^{-8}$ cm^3 s^{-1} [118].

Another aspect that becomes relevant for large tuning ranges is the relative stability of the NIR laser power. According to Eq. 6.21, power fluctuations δP will result in frequency fluctuations $\delta\nu$ of

$$\delta\nu = \frac{\Delta\nu}{2}\frac{\delta P}{P}. \tag{6.22}$$

Achieving 5 MHz resolution over a 50 GHz tuning range requires hence a relative power stability of 2×10^4 within the acquisition bandwidth, while the same resolution for a 5 GHz tuning range can be obtained already for a ten times worse stability level. Since the tuning is caused by the optical generation of an electron-hole plasma, the LIFT method is intrinsically fast. Bimolecular recombination times for relevant excitation densities ($n_{eh} > 10^{16}$ cm^{-3}) are in the nanosecond to sub-nanosecond range.

6.2 Experimental setup

All experiments in this chapter are performed with edge-emitting QCLs, for which the laser oscillates in a ridge waveguide structure between two cleaved facets. The QCLs are Fabry-Pérot uncoated devices that have optical access to both end facets. The frequency tuning effects are observed with a multi-mode diode laser emitting up to several watts of optical power at 809 nm (DILAS, I2F2S22-808-SS14.2). The experimental configuration is shown in Fig. 6.4(a) consisting of two parts: frequency tuning by NIR optical excitation and the molecular spectroscopy with an absorption cell and a fast detector.

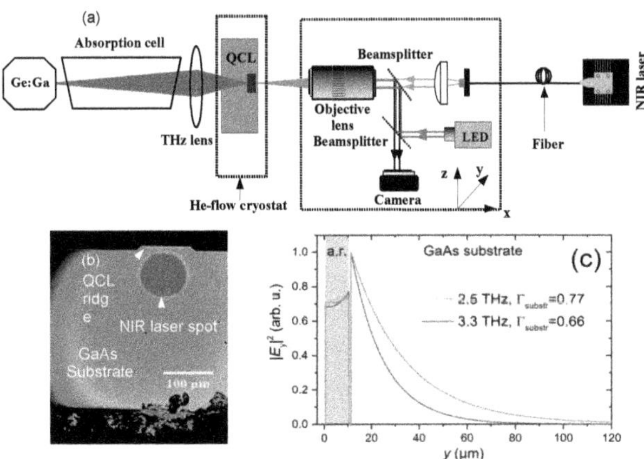

Figure 6.4: (a) The experimental setup for LIFT technique. The QCL is mounted in a He-flow cryostat, (b) Microscope image of the illuminated QCL facet, (c) Calculated profile of the waveguide mode in the vertical (epitaxial-growth) direction for different frequencies (the mode propagates perpendicular to y along the waveguide ridge). The active region (a.r.) has a height of 10 μm and corresponds to the QCL ridge structure in (b).

6.2.1 Setup for frequency tuning

For the tuning setup, the QCL frequency is tuned by illuminating the rear-facet of the QCL with a NIR diode laser. A confocal microscope setup, based on the x-y-z translation stage, is used to align laser beam on the QCL facet. A plano-convex lens with a focal length of 30 mm collimates the laser beam before the objective lens. A small and intense laser spot is generated with an infinity-corrected objective lens (Motic, Plan APO ELWD, 10X) to focus the beam on the QCL facet. QCLs based on single-plasmon waveguides are used for this experiment. For such QCLs, the waveguide mode extends to a large portion of the semi-insulating GaAs substrate. Figure 6.4(b) depicts a microscope image of the illuminated QCL facet showing the substrate with the ridge waveguide structure on top (which contains the active region) and the NIR laser spot. A well-defined tuning behavior is obtained by exciting the GaAs substrate in the vicinity of the ridge waveguide structure. For focused excitation of the active region, a gradual decrease of the QCL intensity, as well as mode instabilities, are observed, which makes such a configuration unfavorable for spectroscopy. Figure 6.4(c) depicts the typical mode intensity profile of a QCL with a single plasmon waveguide. Almost 80% of the mode intensity is guided within the GaAs substrate at 2.5 THz, while at 3.3 THz still two-thirds of the mode intensity is located in the substrate. It is, therefore, rather intuitive that manipulating the optical properties of the region will change the behavior of the QCL. The spot size of the NIR laser is measured to quantify the frequency shift of the THz QCLs with respect to the optical power density. To measure the beam spot, a digital microscope camera (Motic, Moticam 3+) with an active sensing area, 6.55×4.92 mm, is used. During this measurement, the laser was operated slightly higher than its threshold current to prevent any damage to the camera sensor. The captured beam spot was analyzed by the ImageJ processing software.

6.2.2 Setup for molecular spectroscopy

For molecular spectroscopy, the THz emission from of the QCL is collected by a plano-convex TPX lens and directed through a 60 cm long absorption cell onto a Ge:Ga photoconductive detector. Figure 6.4 depicts the setup for the molecular spectroscopy. CH_3OH is used as a molecular species because it has a dense absorption spectrum in the THz frequency region. The QCLs are operated at constant driving current and temperature. The diode laser current is ramped from below threshold to the maximum, and the signal of the Ge:Ga detector is recorded with a fast data acquisition device (CDAQ, National Instruments). Depending on the sweep rate, the acquisition of a single spectrum takes typically 20–100 ms. The maximum illumination intensity for frequency tuning is then determined either by the complete quenching of the QCL emission, an illumination-induced mode transition, or the power limitations of the diode laser.

6.3 Results

The results of a spectroscopic measurement with a QCL operating at 3.3 THz are shown in Figs. 6.5(a). The particular device was chosen here due to the absence of atmospheric absorption characteristics in that frequency range. As a result, the spectral characteristics due to optical excitation are very clearly revealed for this QCL. The short cavity length of 1 mm enables single-mode QCL operation, which is a prerequisite for laser-based spectroscopy. In Fig. 6.5(a), the detector signal is displayed as a function of the driving current of the NIR diode laser. The tuning was achieved by rapid scanning of the diode laser current as well as optical power. The maximum tuning range is restricted by a mode hop for that QCL at a diode laser current of 6.5 A. In the detector signal above the diode laser threshold, a large number of methanol absorption lines are observed along with a gradual decrease in QCL output intensity. Figure 6.5(b) depicts a frequency-calibrated transmission spectrum, for which the data of Fig. 6.5(a) was divided by a reference spectrum for an empty gas cell. For comparison, the simulated transmission spectrum based on the molecular catalog of the JPL is shown [90], which underlines the spectroscopic quality of the experimental data obtained with LIFT. With such a frequency calibration, the tuning behavior can be precisely determined, as shown in Fig. 6.5(c). Where square dots refer to the frequencies of the CH_3OH absorption lines in Fig. 6.5(a), and the solid line is the square-root parameter fit to the data according to Eq. 6.21. As it has been shown in Fig. 6.5(c), the QCL frequency is very well approximated by the square root dependence of the driving current of the diode laser. Since the power of the diode laser scales approximately linearly with the current above the threshold, the frequency change are described in Eq. 6.21. Further experimental results regarding different devices, as well as a discussion of thermal and current tuning parameters, are presented in section 6.5.

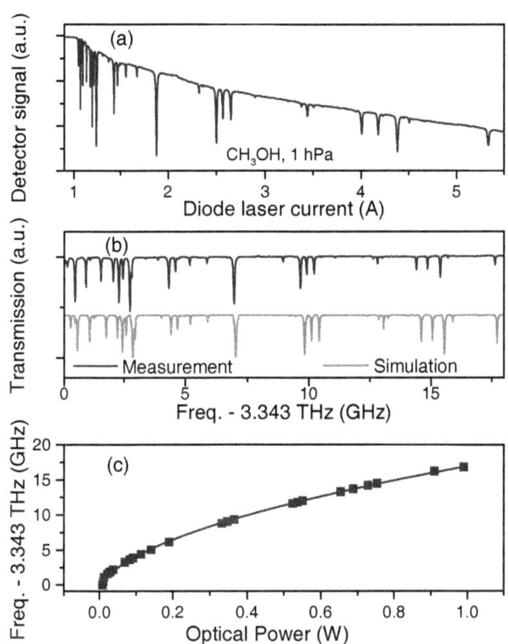

Figure 6.5: (a) Detector signal as a function of the diode laser current for 1 hPa of CH_3OH. The QCL is operated at a constant current of 380 mA at 37 K (threshold value: 300 mA). (b) Measured transmission spectrum after frequency calibration (top) and simulated methanol spectrum (bottom). (c) QCL frequency as a function of the incident diode laser power. Square dots refer to the frequencies of the methanol absorption lines in (a) Solid line: square-root parameter fit to the data according to Eq. 6.21.

6.4 Illumination in different regions of the QCL

As stated in the previous section, a well-defined frequency tuning is observed when the QCL substrate below the active region is illuminated by the NIR diode laser. One measurement is performed here to analyze the tuning behavior due to optical excitation at the different positions of the substrate as well as in the active region. The same measurement setup is used here as described in Fig. 6.4. The measurement is conducted with the 3.1 THz QCL and the frequency of the QCL is tuned by a single-mode diode laser (Lumics, LU0808M250). The signal of the detector after the gas cell is monitored when the NIR laser is illuminated at a different position of the substrate. Figure 6.6(a) indicates the distance used to demonstrate the behavior of the frequency tuning. The distance is measured between the bottom edge of the active region and the top edge of the laser spot is measured. Figures 6.2(b) and 6.2(c) shows the comparison of frequency shifts in the different positions of the substrate. As it is seen, maximum tuning is observed just below the active region of the QCL.

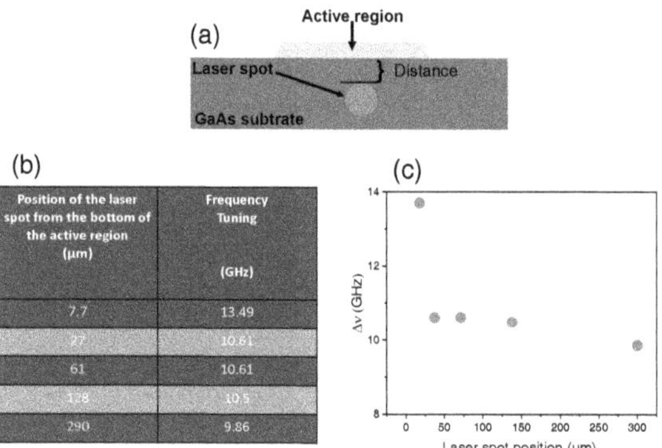

Position of the laser spot from the bottom of the active region (µm)	Frequency Tuning (GHz)
7.7	13.49
27	10.61
61	10.61
128	10.5
290	9.86

Figure 6.6: Frequency tuning behavior due to optical excitation at the different positions of the substrate. (a) Schematic image of the facet of QCL Illuminated by NIR laser. (b) and (c) show the frequency shifts in the different substrate positions.

Figure 6.7 shows the signal of the detector as a function of optical power when the laser excites both the substrate and the active region. The direct excitation in the active region of the QCL exhibits instability in the emission spectra and the spectral lines could not be quantified. The physics for this phenomenon has not yet been understood and will be a subject of future study.

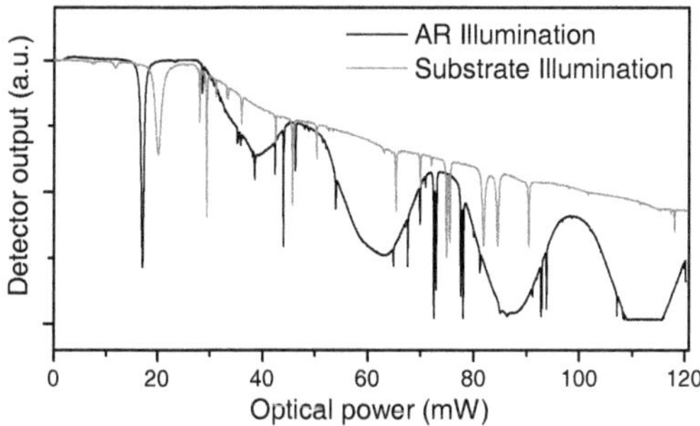

Figure 6.7: Frequency tuning behavior after NIR excitation in the substrate and the active region of the QCL.

6.5 Comparison of light-induced, current and temperature tuning effects

The experiments are successfully performed with QCLs emitting at 2.5, 3.1, and 3.3 THz. All QCL's used for this experiment are Fabry-Pérot lasers with single-plasmon waveguides. The experimental procedures for the characterization of current and temperature tuning are explained in Sect. 4.6. The frequency can be tuned within 1.3–3.8 GHz by changing the QCL driving current, where the smallest and largest tuning ranges are obtained for the 2.5 and 3.1 THz devices, respectively. For temperature tuning, the frequency can be tuned within a range of 1.8–3.5 GHz. An overview of the current and temperature tuning parameters as well as LIFT coverage for each QCL is given in Table 6.1. Since, an increase in temperature leads to a negative frequency shift, while optical excitation leads to a positive shift in frequency, it is clear from this observation that the observed frequency shift is not dominated by heating, but rather by the effect of the LIFT. A frequency shift due to the optically induced heating of the QCL might play a compensating role. However, this effect has to be minor, since even the maximum thermal frequency shift over the whole operating temperature range of the QCLs is still about one order of magnitude smaller than the optically induced frequency shift.

Table 6.1: Current and temperature tuning parameters for the investigated QCLs: maximum frequency coverage $\Delta\nu_{I_{QCL}}$ for a variation of the QCL current (I_{QCL}) and $\Delta\nu_T$ for a variation of the temperature (T), tuning coefficient $d\nu/dI_{QCL}$ for frequency and $d\nu/dT$ for temperature. The last column contains the maximum LIFT range $\Delta\nu_{LIFT}$. The active region design of the corresponding QCLs are given in the respective references.

QCL (THz)	$\Delta\nu_{I_{QCL}}$ (GHz)	$d\nu/dI_{QCL}$ (MHz/mA)	ν_T (GHz)	$d\nu/dT$ (MHz/K)	$\Delta\nu_{LIFT}$ (GHz)
2.5	1.3	+30	1.8	-103	11.4
3.1	3.8	+39	5	-238	39.4
3.3	2.3	+19	3.5	-172	16.8

In order to monitor the frequency coverage of the different QCL modes, LIFT measurements were conducted at different QCL currents. At higher currents, the QCL can emit in different longitudinal modes of the Fabry-Pérot resonator. With the help of molecular spectroscopy, all modes are spectrally resolved by the Ge:Ga detector. Here, the detector signals are shown in Figs. 6.8(a) and 6.8(b) of the 2.5 THz QCL, which was chosen exemplarily. The signals are shown as a function of the laser diode current measured at QCL currents of 405 and 416 mA. Figures 6.8(c) and 6.8(d) show the transmission spectra after applying the baseline correction and calibration technique as described in Sect. 6.3. Single-mode emission is observed at 2.49 and 2.52 THz, with LIFT ranges of 9.4 and 11.4 GHz. This compares to tuning ranges of 1.3 and 1.8 GHz as achieved by current and temperature tuning, respectively as shown in Table 6.1. Figures 6.8(e) and 6.8(f) show the frequency tuning approximation by the square root dependence of the optical power of the diode laser. Where square dots refer to the frequencies of the CH_3OH absorption lines, and the solid line is the square-root parameter fit to the data according to Eq. 6.21.

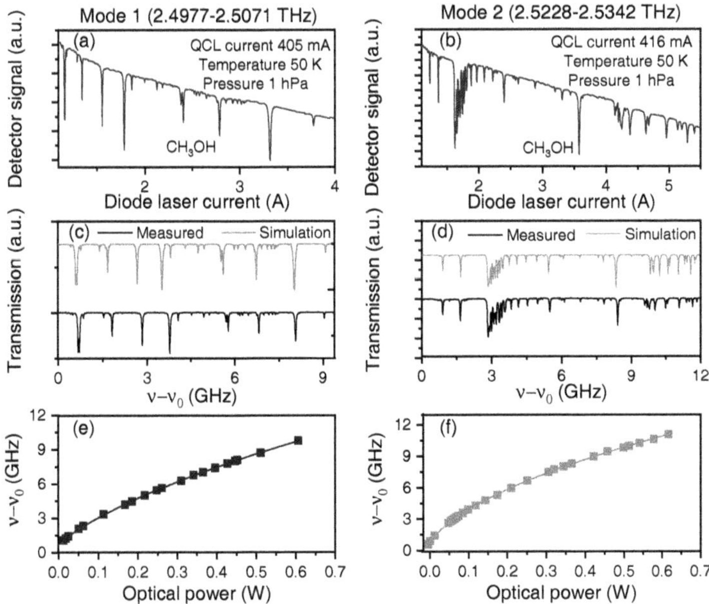

Figure 6.8: Experimental results with a 2.5 THz QCL measured at a temperature of 50 k and a pressure of 1 hPa. The results for mode 1(ν_0= 2.4977 THz) are shown on the left side of the figures and the results for mode 2 (ν_0= 2.5228 THz) are shown on the right.

Table 6.2 summarizes the results obtained for the three QCLs. With the exception of the 3.3 THz QCL, the maximum tuning range was limited by longitudinal-mode transitions of the QCLs. The tuning behavior is well predicted by the one-dimensional model described in Sect. 6.1, despite small variations of the tuning coefficient a_L for the different modes of each QCL. For comparison, the theoretical values for a_L, which correspond to frequency and length of lasers 2.5, 3.1, and 3.3 THz according to Eq. 6.20 are 9.5, 18.2, and 10.9 GHz/$\sqrt{W}$, respectively.

Table 6.2: Single-mode frequency coverage by LIFT for the three QCLs.

QCL (THz)	L (mm)	$\nu_{min} - \nu_{max}$ (THz)	ν_{LIFT} (GHz)	P_{max} (W)	Experimental a_L (GHz/ $\sqrt{W}$)	Theoretical a_L (GHz/ $\sqrt{W}$)
2.5	1.52	2.4977-2.5071	9.4	0.6	16.8	
		2.5228-2.5342	11.4	0.6	15.1	9.5
3.1	0.66	3.091-3.125	33.9	2.2	24.6	
		3.147-3.186	39.4	3.3	23.5	18.2
3.3	1	3.344-3.360	16.8	1	19.3	
		3.382-3.396	14.2	0.9	16.7	10.9

The largest optical tuning range was observed for a 660–μm–long QCL emitting at 3.1 THz, which was illuminated with up to 3.3 W (approximately 50 kW/cm^2). The QCL exhibits single-mode operation in one of two modes at 3.10 THz and 3.15 THz, with LIFT tuning ranges of about 34 and 40 GHz, respectively. Figure 6.9(a) depicts the CH_3OH transmission spectrum for the high-frequency mode of the ~3.1 THz QCL. The LIFT range of almost 40 GHz has been achieved, which is technically limited by the available power of the diode laser. Figure 6.9(b) shows the normalized intensity for 2.5, 3.1 and 3.3 THz QCLs as a function of the normalized optical power. The QCLs intensity decrease with increasing excitation power. This experimentally observed gradual decrease of the QCL output power for larger NIR illumination powers by local heating of the active region, which results in a reduction of the gain, and a decrease of the cavity Q factor, which in turns increases the threshold gain. In contrast to the shift of the main cavity resonance, the decline of the Q factor cannot be easily recovered in the one-dimensional model.

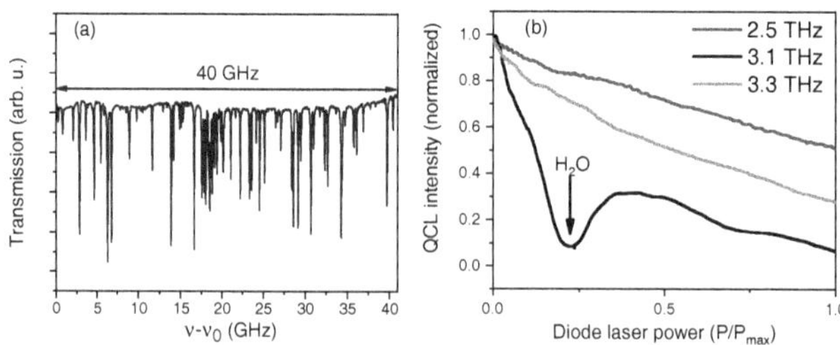

Figure 6.9: (a) CH_3OH transmission spectrum as obtained with a 0.66 mm long 3.1 THz QCL operated at 40 K and 295 mA, $\nu_0 = 3.147$ THz (b) Normalized QCL intensity as a function of the normalized NIR power. The dip marked with an arrow in the characteristics of 3.1 THz QCL is due to atmospheric water absorption.

6.6 Frequency tuning by continuously variable natural density filter

As discussed above, LIFT is obtained by illuminating the GaAs substrate above the bandgap. In order to compare near-surface absorption of GaAs in different wavelengths, measurements are taken with two different excitation sources (Single-mode diode lasers operating at 820 nm and 809 nm). In particular, the 820 nm laser is chosen to increase the penetration depth of the GaAs, as this wavelength corresponds to the GaAs bandgap at low temperatures [118]. The measurement setup used for this experiment is the same as Fig. 6.4(a). Here, however, a natural density (ND) filter (Thorlabs, NDC–50C–4–B) is used between the QCL and the microscope-objective lens. For the spectroscopic measurement, the QCL and the diode laser are operated at constant temperature and current. The optical power of the diode laser is tuned by the continuous variable ND filter. The filter provides a linear attenuation by rotation with an optical density (OD) ranging from 0.04 to 4. The OD is defined as the negative of the logarithm of the transmission ($\mathrm{OD} = -\log_{10}(T)$), where the T varies from 0 to 1. A rotation frequency of 30 Hz is applied to the ND filter, and the signal is obtained by the Ge:Ga detector from the QCL.

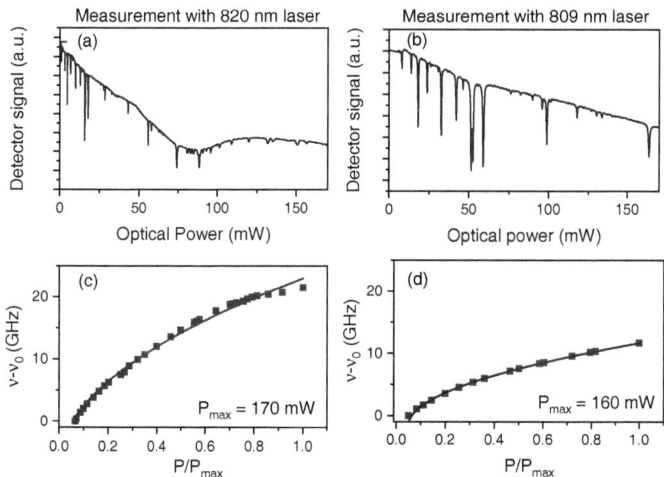

Figure 6.10: Detector signal of CH_3OH as a function of the diode laser intensity. Measured respectively with (a) 820 nm and (b) 809 nm laser diode. Figures (c)–(d) depict the derived frequency (ν) tuning characteristics as a function of the normalized output power, where the solid lines refer to a square-root data fit, and ν_0= 3.14812 THz.

The results of the spectroscopic measurement with a 3.1 THz QCL are shown in the Figs. 6.10(a)–(d). Figures 6.10(a) and 6.10(b) show the transmission signal of CH_3OH as a function of the optical power. If the two spectra from Figs. 6.10(a) and 6.10 (b) are compared to each other, the signal acquired with 809 nm laser has a better signal to noise ratio than the 820 nm laser. That is because the LIFT depends on the relative stability of the NIR laser power. According to Eq. 6.22, the power fluctuation of the NIR laser may result in a direct fluctuation of the QCL frequency. To determine the frequency range, the transmission signals from Figs. 6.10(a) and 6.10(b) are compared to the JPL molecular catalog [90], and the frequency tuning characteristics are plotted as a function of the normalized power in Figs. 6.10(c) and 6.10(d), respectively. As can be seen, the frequency coverage of $\sim$22 GHz is achieved for the 820 nm laser, and for the 809 nm laser, the frequency coverage is approximately 12 GHz with the same level of optical power. The reason behind this broad range of tuning for the 820 nm laser is that the penetration depth is larger at that wavelength in GaAs, resulting in a larger interaction volume. It is clear from this measurement that the frequency coverage of the QCL can be improved in combination with an appropriate wavelength and a power stable diode laser.

6.7 Future prospects

As single-mode operation is facilitated for shorter cavities in combination with a strong LIFT effect [cf. Eq. 6.19], a further decrease of the cavity length may increase the tuning range. A combined bandwidth of 74 GHz is already achieved for the 3.1 THz QCL, and the future milestone will be a LIFT range that exceeds the longitudinal mode spacing. The total continuous frequency coverage will then be determined by mode spacing times the number of modes for which single-mode operation is feasible. As shown in Ref. [119], the gain bandwidth of the active region structure can easily cover several hundred GHz to 1 THz for particular broadband design. In order to reduce the relative power noise of the NIR laser and to keep the frequency fluctuations sufficiently small (cf. Eq. 6.21) additional noise reduction measures will be necessary for a higher optical power. While the presented results are all based on the configuration shown in Fig. 6.4, the next chapter shows a direct coupling to an optical fiber that makes this configuration robust and compact. In such an arrangement, the QCL can be operated inside a mechanical cryocooler without compromising spectroscopic performance, as the cooler vibration does not cause the fiber to displace. Some limitations of the method toward higher frequencies are expected. As illustrated in Subsect. 6.2.1, the waveguide mode becomes stronger confined to the active region with increasing frequency. While almost 80% of the optical mode is found in the substrate for QCLs at 2.5 THz, this number decreases below 50% above 5 THz. If the mode is then confined mainly in the active region, a change in the optical length of the substrate will only act as a small perturbation of the main cavity resonance. Hence, there might be a critical confinement factor, above which large-range optical tuning by substrate excitation becomes impracticable. Unfortunately, direct excitation of the active region often tends to result in unstable operation and can not be easily exploited for spectroscopy. But even if the tuning covers only several hundred MHz, it might be still working perfectly fine for frequency stabilization with a phase-locked loop. In particular, if both output power and frequency stabilization are required, optical tuning might act as a further degree of freedom in addition to temperature and current tuning and ease the stabilization task significantly. The next chapter explains this method for frequency and output power stabilization in more detail.

Chapter 7

Frequency and output power stabilization of a THz QCL using near-infrared illumination

In the heterodyne spectroscopic system, frequency resolution and radiometric accuracy are largely dependent on the frequency and output power stability of the local oscillator (LO). Increasing the stability of the LO with respect to frequency and output power has a direct impact on the performance of a heterodyne receiver. Recent studies have shown that the intrinsic quantum limit for the linewidth of THz QCLs lies in the sub-kHz range [89, 120]. However, the practical linewidth for a free-running THz QCL is much larger, which is due to the limited stability of the driving current and temperature as well as to optical feedback [121]. In practice, the emission frequency drift can exceed 10 MHz [122, 123]. Stabilizing the frequency and output power becomes even more challenging when the QCL is operated in a mechanical cryocooler, where vibration noise is intrinsically present. For these reasons, there has been a significant interest in improving both, the output power and frequency stability. To date, several techniques have been demonstrated for frequency stabilization. Among them are phase locking to a frequency multiplied microwave source [124, 125, 126] or a THz frequency comb [127, 128], locking to a THz gas laser [129, 130], and injection locking of a QCL to a telecommunication frequency comb [131]. An alternative method, which has been well established for diode lasers, is the stabilization to an atomic or molecular absorption line to achieve absolute long-term stability of the laser frequency [132, 133]. This approach has also been demonstrated for QCLs [134, 135]. For a output power control of a frequency-stabilized THz QCL, a swing-arm voice coil actuator has been used as a fast optical attenuator [136]. Recently, amplitude stabilization of a 2.85 THz QCL has been achieved with a graphene-loaded, split-ring-resonator array acting as an external amplitude modulator. In this case, the transmittance of the modulator was controlled by modifying the graphene conductivity via electrostatic back gating [137]. In this chapter, a method will be described and discussed that allows for a simultaneous stabilization of the frequency and output power by taking advantage of the frequency and power regulation by near-infrared (NIR) excitation [2, 113]. Frequency and output power are controlled by modifying the driving currents of the diode laser or the QCL so that no external THz optical modulator is required.

7.1 Experimental setup

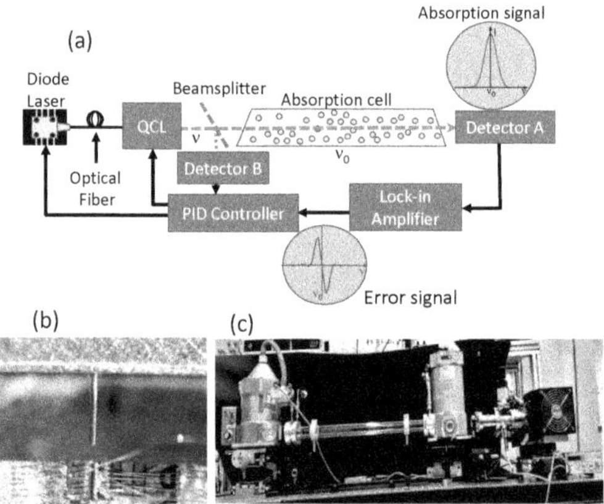

Figure 7.1: (a) The schematic configuration; (b) the photograph of the lab setup used for the frequency and output power stabilization. The QCL is operated in a mechanical cryocooler. The combination of the absorption cell and the Ge:Ga detector (detector A) is used to lock the frequency, and the second Ge:Ga detector (detector B) is used as a reference for the output power stabilization. (c) Rear-facet illumination with a low-NA single-mode fiber.

The experimental configuration and the photograph of the laboratory setup is shown in Figs. 7.1(a) and 7.1(b). The QCL used in this experiment is based on a GaAs/$Al_{0.18}Ga_{0.82}As$ active-region structure [53] with a single-plasmon waveguide. It is an uncoated Fabry-Pérot device, 120 μm wide and 660 μm long, and exhibits single-mode emission at 3.1 THz. The same laser is used for the demonstration of molecular spectroscopy by light-induced frequency tuning in the last chapter. The QCL is operated in a mechanical cryocooler (Ricor, model K535), and the temperature is stabilized at 45 K. The exit window of the cooler is made of high-density polyethylene (HDPE). A TPX lens collimates the QCL beam, and a Mylar beam splitter divides the beam into two parts, where the reflected beam is used for power stabilization, while the transmitted beam passes through a methanol (CH_3OH) gas cell for frequency locking to a molecular absorption line. Two helium-cooled Ge:Ga detectors (labeled A and B) are used to measure the transmitted and the reflected signals. A single-mode NIR diode laser (Lumics, LU0808M250) is used as a source for illuminating the rear facet of the THz QCL. The diode laser emits 200 mW of optical power into a polarization-maintaining single-mode fiber with a core diameter of 4.5 μm and a numerical aperture (NA) of 0.12. The fiber is clamped to a linear stepper positioner (Attocube, ANPx51) driven by a motion controller (Attocube, ANC 300) to allow for the lateral alignment of the NIR spot on the QCL chip. Since the fiber is attached to the cold finger as shown in Fig. 7.1(b), cooler vibrations do not have an impact on the alignment as compared to our microscope technique described in [2]. The DC current for the QCL is provided by the driver QCL1000 LAB (Wavelength Electronics). The driver has a typical modulation bandwidth of up to 3 MHz and a modulation coefficient of 200 mA/V.

7.2 Frequency tuning measurements

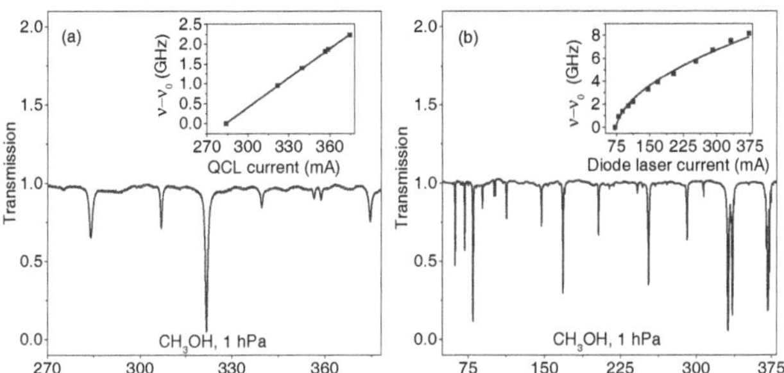

Figure 7.2: Absorption spectra of CH_3OH in a 60-cm cell, at a pressure of 1 hPa and heat sink temperature of 45 K, (a) by varying the QCL driving current and (b) by varying the diode laser intensity. The insets in (a) and (b) depict the derived frequency (ν) tuning characteristics, where the solid lines refer to a linear and a square-root data fit, respectively, and $\nu_0 = 3.14589$ THz.

In order to identify a frequency range suitable for stabilization, the absorption characteristics of CH_3OH are investigated using two different tuning methods. The absorption spectrum was first measured as a function of the QCL current from 250 to 380 mA at a temperature of 45 K, while the NIR diode laser was switched off. Figure 7.2(a) shows the resulting spectrum as a function of the driving current exhibiting seven absorption lines. Second, light-induced frequency tuning was performed, for which the QCL is operated at a constant driving current and the diode laser current is ramped from below threshold up to 370 mA. The corresponding absorption signal after a fourth-order polynomial baseline subtraction is shown in Fig. 7.2(b). The individual lines of the CH_3OH spectral fingerprint are identified with the help of the JPL molecular catalog [90]. With such a frequency calibration, the tuning behavior can be precisely recovered. By changing the QCL driving current, a frequency coverage of about 2.5 GHz [cf. inset of Fig. 7.2(a)] is achieved. To a reasonable approximation, the QCL frequency depends linearly on the QCL current as well as on the heat sink temperature (not shown). For the used laser, we identified a suitable single-mode frequency range of 3.1455–3.148 THz. In the case of diode laser tuning, the QCL frequency shift follows a square root behavior with an increase in diode laser current above threshold. The details of this method have been discussed in the previous chapter and can also be found in [2]. In this tuning mode, a frequency coverage of about 8 GHz [cf. inset of Fig. 7.2(b)] was achieved with only 200 mW of optical power.

7.3 Frequency modulation spectroscopy

In order to lock the frequency to a molecular absorption line, a distinctive signal is required for the control loop, which contains information about the magnitude and the direction of the frequency deviation. Therefore, a small sinusoidal modulation at 1 MHz is added to the QCL or diode laser driving current and the detector signal after the absorption cell is recorded with a lock-in amplifier (Zurich Instruments, UHFLI). This frequency modulation (1f) results in a derivative-like absorption signal. At the same time, it provides phase and amplitude information corresponding to the direction and magnitude, respectively, of the required corrections. Figures 7.3(a) and 7.3(b) show the derivative-like absorption spectrum of CH_3OH at a pressure of 1.5 hPa measured at heat sink temperatures of 45 K and 43 K, respectively. For the experimental result shown in Fig. 7.3(a), the QCL current is tuned from below threshold over the entire range of single-mode operation and for Fig. 7.3(b), the QCL operates at a constant current of 275 mA and the laser diode current increases from below the threshold to 370 mA resulting in an acquisition time of 50–100 ms/scan. Each line shown in Figs. 7.3(a) and 7.3(b) can be used to provide an error signal for the QCL frequency locking. All signals are acquired with a fast data acquisition device (CDAQ, National Instruments), where both, the input and output channels, are synchronized with the lock-in amplifier at a clock rate of 10 MHz.

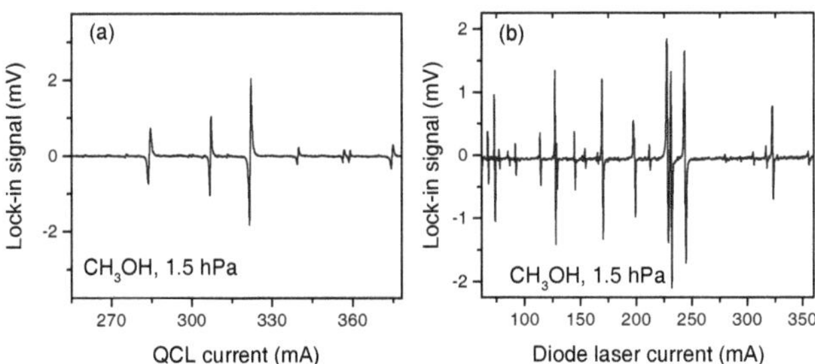

Figure 7.3: Derivative-like absorption spectrum of CH_3OH for a modulation frequency of 1 MHz, a cell length of 60 cm, and at a pressure of 1.5 hPa (a) QCL modulation (b) Diode laser modulation

7.4 Frequency and output power stabilization

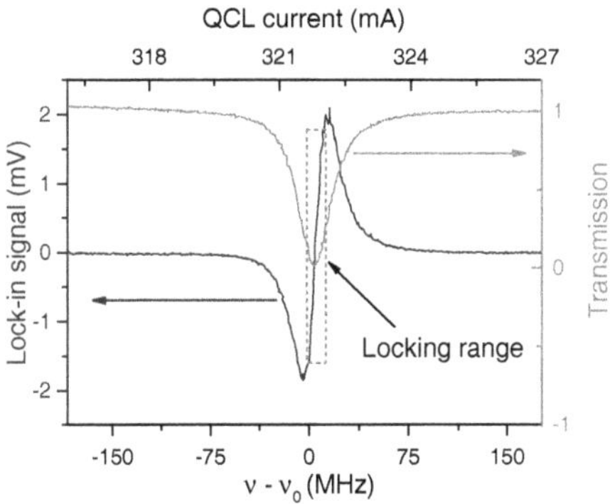

Figure 7.4: Absorption line of CH_3OH for a pressure of 1.5 hPa, where $\nu_0 = 3.14685$ THz. The transmission signal is measured in direct detection, while the frequency-modulated signal is measured with the lock-in amplifier. The locking range is indicated by the rectangular box.

Figure 7.4 shows the CH_3OH absorption line at 3.14685 THz and the corresponding derivative-like signal, which is used for the frequency stabilization. The absorption signal is measured directly with the Ge:Ga detector through a variation of the QCL driving current at a pressure of 1.5 hPa and the frequency-modulated signal is measured with the lock-in amplifier. The derivative-like signal changes linearly with the QCL driving current close to the center of the absorption line. Therefore, the error signal can be converted directly into a frequency scale. The linear range of the error signal also indicates the locking range of the control loop.

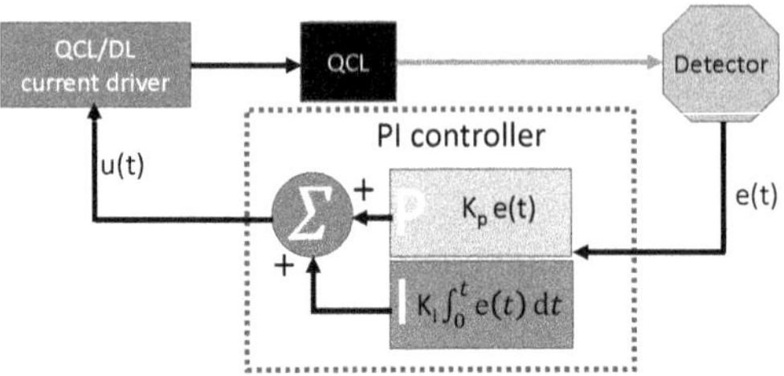

Figure 7.5: Schematic block diagram of the PI control loop for the laser stabilization system.

The lock-in amplifier (UHFLI, Zurich Instruments) used in this measurement has an integrated programmable PID feature. It consists of four configurable PID (proportional-integral-derivative) controllers with an autotuned PID feature. Figure 7.5 shows the schematic block diagram of the PI control loop for the laser stabilization system. Two control parameters are mainly used in this locking approach: proportional to the error itself (P), proportional to the integral of the error (I). The overall control function is

$$u(t) = K_p e(t) + \frac{1}{T_i} K_p \int e(t)dt = K_p e(t) + K_i \int e(t)dt. \tag{7.1}$$

The proportional gain (K_p) corrects the error, and the integral part (K_i) accumulates the error over integration time T_i and is used to correct a possible residual offset. The controller provides feedback to the QCL driving current to keep its frequency aligned with the center of the absorption line. The noise level and time constant of the system resulted in optimal parameters for the proportional gain of K_p=50 and for the integral gain of K_i=400.

A second feedback loop is used to achieve amplitude stabilization by varying the diode laser driving current, where the Ge:Ga detector B signal is used as a reference for the QCL power level. The deviation of the power signal from a predefined setpoint is used as an error signal. By combining both control loops, the QCL frequency and power level can be stabilized without the need for an external THz modulator. Here, the optimum values for K_p and K_i are -0.1 and -100 k, respectively. To characterize the stabilization approach, the fluctuations of the frequency and output power of the QCL were monitored for different cases. Figures 7.6(a) and 7.6(b) show the frequency and output power stability, respectively, for (1) the free-running QCL, (2) the frequency-locked QCL, and (3) the frequency and power-locked QCL. In the free-running case, the rather strong frequency and amplitude fluctuations are correlated with the cooler vibrations (45-Hz cycle of the piston). In the second case when the frequency stabilization loop is enabled, the frequency becomes well stabilized, but fluctuations in the QCL output power increase. In the third case, when both control loops are activated simultaneously, the power and frequency remain stable at the expense of a very small increase in the frequency noise level. In Fig. 7.6(c), the noise signal from detector A is shown together with the error signal for the stabilized QCL. The noise level was measured by blocking the QCL radiation in front of the detector. This illustrates that the QCL frequency stabilization operates already close to the detector noise limit.

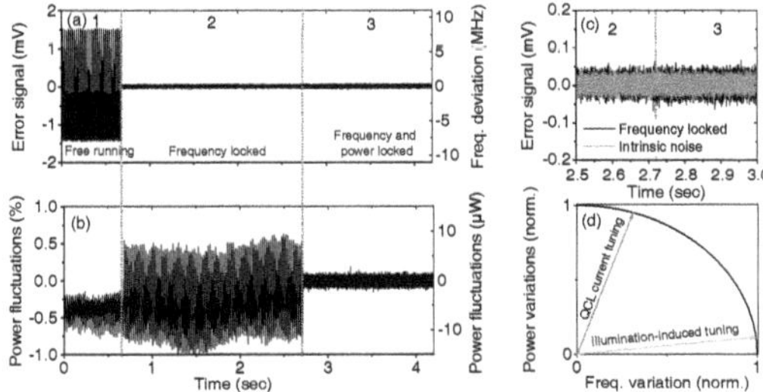

Figure 7.6: (a) Frequency stability versus time detected by the lock-in amplifier in the locking range indicated in Fig. 7.4. (b) DC output of detector B reflecting the QCL output power versus time. Three different cases are shown: (1) free running, (2) frequency stabilized, and (3) both, frequency and amplitude, stabilized. (c) Magnified region around 2.7 sec showing the frequency-locked signal in comparison to the detector noise level. (d) Normalized tuning vectors for the QCL current and the diode laser tuning.

In order to explain the anti-correlation between the frequency and the output power stability in the second case, the impact of the control parameters on the QCL behavior needs to be discussed. As discussed in the last chapter, the QCL frequency and the output power depend on three control parameters: the QCL driving current, the heat sink temperature, and the diode laser excitation power. The temperature control is typically too slow for an active stabilization. Since the control parameters simultaneously affect the QCL frequency and output power, stabilizing one quantity will result in increased fluctuations of the other one. Simultaneous frequency and power stabilization therefore requires two separate control loops operating at the same speed. Another requirement is that the ratio of the power-to-frequency tuning must differ for the two control parameters. This is in analogy to the requirement of two linear independent base vectors to span a two-dimensional phase space. With the help of the tuning coefficients, we can define a pair of tuning vectors $\boldsymbol{\alpha}_I$ and $\boldsymbol{\alpha}_L$ for current and illumination tuning, respectively, at the particular operating point chosen in Fig. 7.6 for stabilization. For QCL current tuning, $\boldsymbol{\alpha}_I = (+24.65$ MHz/mA, $+13.7$ μW/mA), while for illumination tuning $\boldsymbol{\alpha}_L =$ ($+59.4$ MHz/mA, -1.25 μW/mA). Eventually the frequency and power variations are described by the equation

$$\begin{pmatrix} \Delta\nu \\ \Delta P \end{pmatrix} = \Delta I_{\mathrm{QCL}}\,\alpha_I + \Delta I_{\mathrm{DL}}\,\alpha_L\,. \tag{7.2}$$

In order to compare the vector for current tuning with the one for illumination tuning, the tuning vectors are divided by the free-running root-mean-square (rms) fluctuations of 6.2 MHz and 1.6 μW, respectively, with the latter one corresponding to about 0.1% of the QCL power of 1.5 mW. The dimensionless tuning vectors are eventually obtained by normalizing their length to unity, because the actual value of the control parameter is not important here. The resulting normalized tuning vectors are shown in Fig. 7.6(d). In the present case, they span an angle of about 65° in the corresponding normalized phase space. For minimizing crosstalk effects, the tuning vectors should be ideally orthogonal and aligned along the frequency and power axis. In theory, such a pair of orthogonal tuning vectors can be obtained by a simple base transformation. A real-time implementation of such an improved PID loop may be realized with the help of a controller based on a field-programmable gate array.

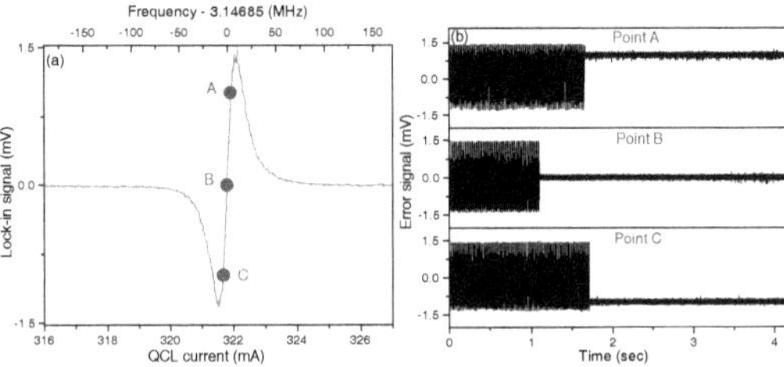

Figure 7.7: Dependence of the output frequency of the THz QCL by changing the set-point at the different point of the absorption line through the PID software system.

Another experiment was carried out to check whether or not the QCL is locked to a molecular resonance by changing the set-point of the PID controller. This experiment was carried out not only to study the linewidth of the QCL frequency but also to adjust the set frequency value within certain limits without the deactivation of the stabilization system. Figure 7.7 illustrates the frequency stabilization at different points of the absorption line. Initially, the frequency was not locked to the molecular resonance, hence the error signal corresponds to the fluctuation of the QCL output frequency in free-running mode. After that, the frequency stabilization system is engaged with different set points (A, B, and C). Figure 7.7(b) demonstrates that the QCL output frequency can be controlled in a range of several MHz around the center of the molecular absorption line. From this measurement, it is clear that the tunable THz QCL with a frequency stabilization system based on a fast detection scheme presents new and extensive possibilities for real-time control of the laser frequency within the continuous frequency range of the laser.

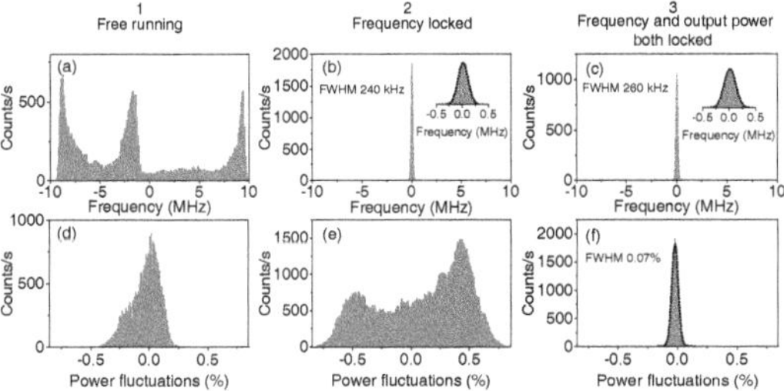

Figure 7.8: Histograms of the (a)–(c) frequency and (d)–(f) power fluctuations corresponding to the unlocked state and the locked states shown in Fig. 7.4. Dotted lines refer to a normal distribution.

Figure 7.8 depicts the frequency and output power distributions for the free-running and stabilized cases shown in Fig. 7.6. Figure 7.8(a) depicts a histogram of the frequency fluctuations for the free-running state. The distribution spans almost 20 MHz, and its non-Gaussian shape is a consequence of the periodic cooler vibrations. The width of the distribution has to be considered as a lower limit since it exceeds the linear range of the error signal. For the frequency-locked state shown in Fig. 7.8(b), the distribution exhibits a Gaussian line shape with a full width at half maximum (FWHM) of 240 kHz, and, for the fully locked state, the line shape corresponds to a FWHM of 260 kHz shown in Fig. 7.8(c). For comparison, the noise level of detector A as shown in Fig. 7.6(c) corresponds to a Gaussian line shape with a FWHM of 200 kHz. These values indicate that the intrinsic noise level of the detector contributes significantly to the linewidth of the stabilized QCL. Figures 7.8(d), 7.8(e), and 7.8(f) compare the power stability for the cases of the free-running, frequency-locked, and fully stabilized QCL, respectively. The fluctuations amount to 0.1% (rms) in the free-running case. For the frequency-locked case without power stabilization, the output power fluctuations of the QCL increased to about 0.4% (rms). When both, the amplitude and frequency stabilization loops are active, the output power becomes well stabilized, and the fluctuations are reduced to 0.03% (rms), corresponding to a Gaussian distribution with a FWHM of 0.07%. This is a more than three-fold improvement with respect to the unstabilized case. The frequency of the THz QCL can also be stabilized by regulating the current of the laser diode. In that case, the power is stabilized by modifying the driving current of the QCL. The results of this method are shown in section 7.5.

In order to illustrate the impact of the cooler vibrations and the control loop action, Figs. 7.9(a) and 7.9(b) depict the power spectral density (PSD) for the frequency and the amplitude noise, respectively, for the three situations shown in Fig. 7.6. The noise PSD corresponds to the magnitude squared of the Fourier-transformed frequency and amplitude fluctuations. Note that the square root of the integrated PSD must be equal to the rms standard deviation of the corresponding distribution. For a Gaussian distribution, this is equal to the FWHM value divided by 2.355. For the free-running case, the PSDs of the frequency and amplitude noise are clearly dominated by the cooler vibrations at 45 Hz and higher harmonics. These contributions become suppressed in the frequency noise as soon as the frequency stabilization loop is activated. However, they become more pronounced in the amplitude noise due to a significant power component of the tuning vector. After activating the second control loop, they are also suppressed in the PSD of the amplitude noise at the expense of a somewhat increased noise at high frequencies, which we attribute to the crosstalk between the control loops by the non-orthogonal tuning vectors.

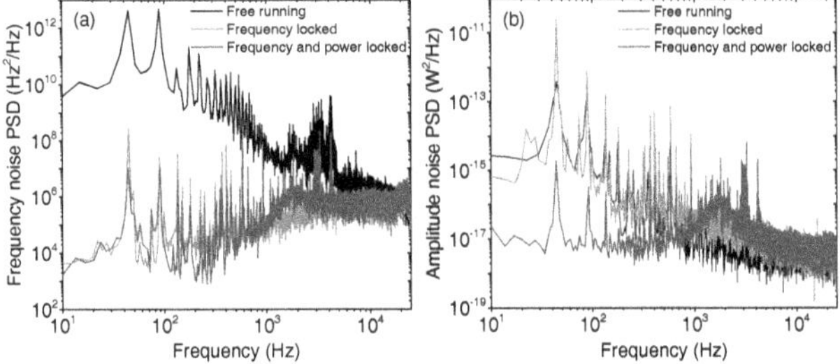

Figure 7.9: PSD of the (a) frequency noise and (b) amplitude noise for the free-running, the frequency-locked, and the frequency and power-locked QCL. The bandwidth of the PID control loops for frequency and output power stabilization is 193 kHz and 1.7 kHz, respectively.

7.5 Frequency stabilization with NIR laser and power stabilization with QCL driving current

This is an alternative method for stabilizing THz QCLs. In this case, the QCL frequency was locked to the molecular absorption line by means of the NIR laser, and the output power was stabilized by the QCL driving current. The same experimental setup was used here, as shown in 7.1. In this case, a small sinusoidal modulation at 1.5 MHz is applied to the diode laser current, and both the diode laser and the QCL are operated at constant driving current. The results for this type of stabilization are shown in Figs. 7.10

and 7.11. For the characterization of the stabilization, the fluctuations were monitored as in section 7.4 for three different cases. Figs. 7.10(a) and 7.10(b) show the frequency and output power stability for (1) the free-running QCL, (2) the frequency locked QCL, and (3) the frequency and power locked QCL.

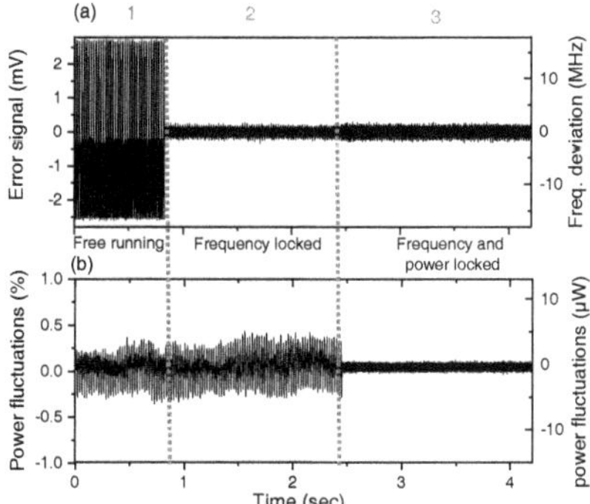

Figure 7.10: Frequency stability versus time detected by the lock-in amplifier in the locking range indicated in Fig. 7.4. (b) DC output of detector B reflecting the QCL output power versus time. Three different cases are shown: (1) free running, (2) frequency stabilized, and (3) both, frequency and amplitude, stabilized.

For statistical quantification of frequency and power stabilization, all results shown in Fig. 7.10 are displayed as normal distributions in Fig. 7.11. Figures 7.11(a), 7.11(b), and 7.11(c) illustrate the frequency distributions for the free-running and stabilized cases. However, compared with the results by using the QCL current for the frequency locking to the previous section, the distribution exhibits a larger FWHM of 550 kHz. For the fully locked state, the line shape corresponds to an FWHM of 700 kHz, which is more than twice the value that of the previous approach used for frequency stabilization. Figures 7.11(d), 7.11(e), and 7.11(f) compare the power stability for the cases of the free-running, frequency-locked, and fully stabilized QCL, respectively. Compared to the previous method, the amount of power fluctuations in the free-running case is almost the same. The output power variations of QCL increases to about 0.15% (rms) in the frequency-locked case. When both the amplitude and frequency stabilization loops are active, the output power becomes well stabilized, and the fluctuations are reduced to 0.017% (rms), corresponding to a Gaussian distribution with an FWHM of 0.037%. In that case, the power stability is improved by a factor of two compared to the previous method.

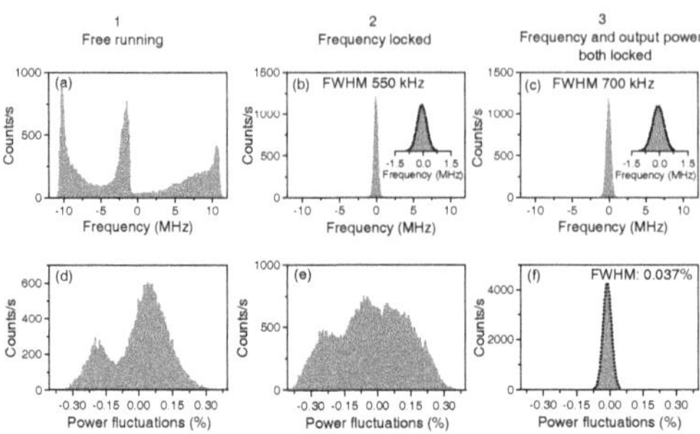

Figure 7.11: Histograms of the (a)–(c) frequency and (d)–(f) power fluctuations corresponding to the unlocked state and the locked states shown in Fig. 7.4. Dotted lines refer to a normal distribution.

Both techniques provide comparable levels of frequency and power stability. Compared to each other, when the focus is on frequency stability, the first approach (section 7.4) performs better and the second approach (section 7.5) does well when focusing on power stability. Overall, the stabilizing frequency or power via the QCL driving current seems to perform better than via diode laser excitation. The reason for that might be a technical one: while the analog bandwidth of the current driver is 3 MHz, the analog bandwidth of the diode laser driver is only 100 kHz. Therefore, performance improvements can be expected to use faster driving electronics for the diode laser. The relative stability of the NIR laser power may have a measurable impact on the QCL stability. As can be seen from the previous chapter, power fluctuations in the laser diode cause a fluctuation in frequency. That might be an additional reason why the same level of output power and frequency stabilization appears to be more challenging to achieve via the diode laser current than via the QCL driving current.

7.6 Longterm measurement

In the previous section, all the locking processes are performed on a shorter time scale. Nevertheless, in many applications, it is essential to have a narrow laser linewidth with low spectral drift over a long period. QCL temperature drifts and bias current fluctuations are the main factors that cause long-term frequency drifts and output power fluctuations. Even for passively stabilized QCLs based on thermal and electrical bias control, linewidth under standard laboratory conditions may exceed several megahertz over longer acquisition times [82]. Long-term measurements are therefore carried out using the above-mentioned method to check the quality of the lock for a certain time scale. In order to perform this measurement, the error signals from detectors A and

B were recorded for a period of 30 minutes undisturbed during the lock. Figs. 7.12(a) and 7.12(b) show the frequency and output power stability, respectively, for free running and the frequency and power locked QCL. In the free-running case, strong frequency and amplitude fluctuations are observed. When the frequency and power stabilization loop are enabled at the same time, the output of both detectors was stabilized. During the whole locking period shown in the Figs. 7.12(a) and 7.12(b), the frequency distribution exhibits a Gaussian line shape with a full width at half maximum (FWHM) of 220 kHz, and the power fluctuations show a Gaussian distribution with an FWHM of 0.1% [cf. insets of Figs. 7.12(a) and 7.12(b)]. The long-term stabilization results differ slightly from those previously measured due to the variance of the PID parameters.

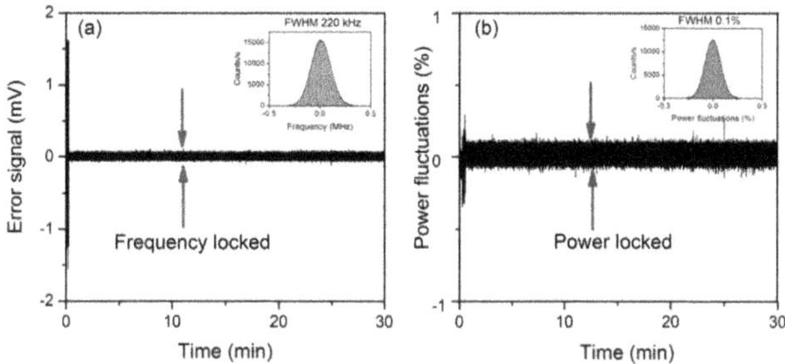

Figure 7.12: Longterm frequency and output power stabilization measurement. The insets in (a) show the distribution of the frequency locked portion which exhibits a Gaussian line shape with a FWHM of 220 kHz and (b) show the distribution of the power locked portion corresponding to the Gaussian distribution of 0.1%

7.7 Conclusion

The simultaneous stabilization of the frequency and output power of a THz QCL is successfully demonstrated in this chapter. Frequency stabilization is achieved by locking the QCL to a CH_3OH absorption line. Frequency and output power are adjusted via the QCL driving current and the power of the NIR laser excitation of the QCL rear facet, which act as two independent control parameters. A long-term linewidth of 260 kHz (full width at half maximum) and a root-mean-square power stability below 0.03% are achieved. With respect to the free-running case, this stabilization scheme improves the frequency stability by nearly two orders of magnitude and the power stability by a factor of three. The setup is compact and robust due to the direct coupling of the illumination through an optical fiber. This scheme could play a vital role in improving the performance of high-resolution spectroscopic systems.

Chapter 8

Summary and outlook

This thesis presented THz quantum cascade lasers (QCLs) as a spectroscopic tool and, in particular, their applications in high-resolution molecular spectroscopy. The main objectives of this thesis were to characterize the tunability and spectral purity of cw THz QCLs using the abundant absorption features of molecules in the THz range and develop methods for high-resolution broadband spectroscopy. In order to achieve these objectives, three different topics have been investigated.

First, a Lamb dip experiment was conducted to study optical saturation and achieve a sub-Doppler resolution for molecular transitions. This work, to my knowledge, is the first investigation of the Lamb dip effect with a THz QCL. The measurements are based on a strong rotational transition of HDO. They confirm the feasibility of QCL-based sub-Doppler spectroscopy, and a sub-MHz emission linewidth of free-running QCLs. In addition, it has been found that a certain level of external optical feedback is tolerable as long as the free spectral range of the external cavity is large compared to the width of the absorption line. The Lamb dip was observed as a function of optical power and pressure, and the lineshape of the Lamb dip was found to be a combination of power, pressure, and transit time broadening. The developed method might be useful for future Lamb dip stabilization of QCLs and Doppler-free precision spectroscopy.

The second focus of this thesis was to improve the frequency tunability of THz QCLs. The development of **L**ight **I**nduced **F**requency **T**uning (LIFT) is a promising approach because it combines a wide tuning range with easy implementation. Using LIFT, the feasibility of high-resolution molecular spectroscopy with wide bandwidth has been demonstrated based on a well-controlled illumination method. This technique exploits the change of the dielectric properties of GaAs when illuminated with light above the bandgap. By applying this method, a tuning range of 40 GHz is obtained for a 3.1 THz QCL, a ten-fold improvement over the usual electrical tuning. A physical model is developed that explains the observed frequency tuning characteristics by optical excitation of an electron-hole plasma. The integration of single-mode fiber to illuminate the QCL substrate allows for a very compact implementation and is compatible with nano-optical bench approaches. LIFT could facilitate broadband molecular spectroscopy that allows for solid/liquid state spectroscopy with THz QCLs.

The third objective was to stabilize the emission frequency and output power of the THz QCL. The performance of sensitive measurements, such as heterodyne spectroscopy, depends not only on the reduction of QCL emissions linewidth down to sub-megahertz levels but also on the output power stability. The method that has been developed here allows for the simultaneous stabilization of frequency and output power by taking advantage of near-infrared (NIR) optical excitation. The QCL frequency was locked to a CH_3OH molecular transition by employing modulation spectroscopy and a PID feedback loop. Starting with a free-running linewidth of more than 20 MHz, the frequency linewidth was reduced to 260 kHz (FWHM) after stabilizing to the molecular absorption line. The power of the QCL is stabilized by the NIR diode laser using a second PID loop with a reference detector. The output power fluctuations of the QCL are reduced from 0.4% to as low as 0.03% (rms). The required rather large absorption cell in present approach might be replaced by a harmonic mixer and a multiplied source for future spaceborne applications.

Although significant challenges remain ahead, THz QCLs are currently the only compact source operating with high output power above 2 THz. Cryogenic cooling is a severe limitation that prevents the routine use of THz QCLs. Frequency stabilization and tunability are the key parameters to be improved along with the operating temperature. In addition, higher average output power is desirable to further increase the sensitivity of spectrometers using QCL THz sources. The ultimate goal will be to develop delicate, frequency-tunable, coherent THz spectrometers using QCLs as transmitters or local oscillators. The techniques presented in this thesis will enable powerful QCL-based high-resolution THz spectroscopy in the near future. Together with a wide tuning range of THz QCL tuning, such a compact spectrometer can be deployed in real-world applications, including environmental monitoring, astronomical observation, non-destructive testing, and many more.

Bibliography

[1] K. T. Smith, "Molecules in interstellar space," *Science*, vol. 363, no. 6428, pp. 704–705, 2019.

[2] T. Alam, M. Wienold, X. Lü, K. Biermann, L. Schrottke, H. Grahn, and H.-W. Hübers, "Wideband, high-resolution terahertz spectroscopy by light-induced frequency tuning of quantum-cascade lasers," *Optics Express*, vol. 27, no. 4, pp. 5420–5432, 2019.

[3] M. Wienold, T. Hagelschuer, N. Rothbart, L. Schrottke, K. Biermann, H. T. Grahn, and H.-W. Hübers, "Real-time terahertz imaging through self-mixing in a quantum-cascade laser," *Applied Physics Letters*, vol. 109, no. 1, p. 011102, 2016.

[4] S. Koenig, D. Lopez-Diaz, J. Antes, F. Boes, R. Henneberger, A. Leuther, A. Tessmann, R. Schmogrow, D. Hillerkuss, R. Palmer, T. Zwick, C. Koos, W. Freude, O. Ambacher, J. Leuthold, and I. Kallfass, "Wireless sub-THz communication system with high data rate," *Nature Photonics*, vol. 7, pp. 977–981, 2013.

[5] P. Doradla, C. Joseph, and R. H. Giles, "Terahertz endoscopic imaging for colorectal cancer detection: Current status and future perspectives," *World Journal of Gastrointestinal Endoscopy*, vol. 9, no. 8, pp. 346–358, 2017.

[6] Terasense, "Security applications." `https://terasense.com/applications/security`, Accesed: 09–01–2020.

[7] NASA, "SOFIA Overview." `https://www.nasa.gov/mission_pages/SOFIA/overview/index.html`, Accesed: 09–01–2020.

[8] T. Torii, H. Chiba, T. Tanabe, and Y. Oyama, "Measurements of glucose concentration in aqueous solutions using reflected THz radiation for applications to a novel sub-THz radiation non-invasive blood sugar measurement method," *Digital Health*, vol. 3, p. 2055207617729534, 2017.

[9] M. I. Lvovska, N. C. Seeman, R. Sha, T. R. Globus, T. B. Khromova, and T. S. Dorofeeva, "THz Characterization of DNA Four-Way Junction and Its Components," *IEEE Transactions on Nanotechnology*, vol. 9, no. 5, pp. 610–617, 2010.

[10] N. Rothbart, O. Holz, R. Koczulla, K. Schmalz, and H.-W. Hübers, "Analysis of Human Breath by Millimeter-Wave/Terahertz Spectroscopy," *Sensors (Basel, Switzerland)*, vol. 19, no. 12, p. 2719, 2019.

[11] M. Borovkova, M. Khodzitsky, P. Demchenko, O. Cherkasova, A. Popov, and I. Meglinski, "Terahertz time-domain spectroscopy for non-invasive assessment of water content in biological samples," *Biomedical Optics Express*, vol. 9, no. 5, pp. 2266–2276, 2018.

[12] G. J. Stacey, "THz Low Resolution Spectroscopy for Astronomy," *IEEE Transactions on Terahertz Science and Technology*, vol. 1, no. 1, pp. 241–255, 2011.

[13] H. Richter, M. Wienold, L. Schrottke, K. Biermann, H. T. Grahn, and H.-W. Hübers, "4.7–THz Local Oscillator for the GREAT Heterodyne Spectrometer on SOFIA," *IEEE Transactions on Terahertz Science and Technology*, vol. 5, no. 4, pp. 539–545, 2015.

[14] I. Kallfass, I. Dan, S. Rey, P. Harati, J. Antes, A. Tessmann, S. Wagner, M. Kuri, R. Weber, H. Massler, *et al.*, "Towards MMIC-based 300GHz indoor wireless communication systems," *IEICE Transactions on Electronics*, vol. 98, no. 12, pp. 1081–1090, 2015.

[15] M. Fujishima, S. Amakawa, K. Takano, K. Katayama, and T. Yoshida, "Tehrahertz CMOS design for low-power and high-speed wireless communication," *IEICE Transactions on Electronics*, vol. 98, no. 12, pp. 1091–1104, 2015.

[16] T. Nagatsuma, S. Horiguchi, Y. Minamikata, Y. Yoshimizu, S. Hisatake, S. Kuwano, N. Yoshimoto, J. Terada, and H. Takahashi, "Terahertz wireless communications based on photonics technologies," *Optics Express*, vol. 21, no. 20, pp. 23736–23747, 2013.

[17] G. P. Gallerano, S. Biedron, *et al.*, "Overview of terahertz radiation sources," in *Proceedings of the 2004 FEL Conference*, vol. 1, pp. 216–221, 2004.

[18] H. Eisele and G. I. Haddad, "Two-terminal millimeter-wave sources," *IEEE Transactions on Microwave Theory and Techniques*, vol. 46, no. 6, pp. 739–746, 1998.

[19] H. Eisele, M. Naftaly, J. Fletcher, D. Steenson, and M. Stone, "The study of harmonic-mode operation of GaAs TUNNETT diodes and InP Gunn devices using a versatile terahertz interferometer," *Spectroscopy*, vol. 11, p. 12, 2004.

[20] M. Y. Glyavin, A. G. Luchinin, and Y. V. Rodin, "Generation of 5 kW/1 THz coherent radiation from pulsed magnetic field gyrotron," in *35th International Conference on Infrared, Millimeter, and Terahertz Waves*, pp. 1–3, IEEE, 2010.

[21] M. Glyavin, G. Denisov, V. Zapevalov, A. Kuftin, A. Luchinin, V. Manuilov, M. Morozkin, A. Sedov, and A. Chirkov, "Terahertz gyrotrons: State of the art and prospects," *Journal of Communications Technology and Electronics*, vol. 59, pp. 792–797, 2014.

[22] D. Saeedkia, *Handbook of terahertz technology for imaging, sensing and communications*. Elsevier, 2013.

[23] E. Bründermann, H.-W. Hübers, and M. F. Kimmitt, *Terahertz Techniques*, vol. 151. Springer Series in Optical Sciences, 2012.

[24] D. T. Hodges, "A review of advances in optically pumped far-infrared lasers," *Infrared Physics*, vol. 18, no. 5-6, pp. 375–384, 1978.

[25] A. Pagies, G. Ducournau, and J.-F. Lampin, "Low-threshold terahertz molecular laser optically pumped by a quantum cascade laser," *APL Photonics*, vol. 1, no. 3, p. 031302, 2016.

[26] M. Mičica, S. Eliet, M. Vanwolleghem, R. Motiyenko, A. Pienkina, L. Margulès, K. Postava, J. Pištora, and J.-F. Lampin, "High-resolution THz gain measurements in optically pumped ammonia," *Optics Express*, vol. 26, no. 16, pp. 21242–21248, 2018.

[27] K. Vijayraghavan, R. W. Adams, A. Vizbaras, M. Jang, C. Grasse, G. Boehm, M. C. Amann, and M. A. Belkin, "Terahertz sources based on Čerenkov difference-frequency generation in quantum cascade lasers," *Applied Physics Letters*, vol. 100, no. 25, p. 251104, 2012.

[28] A. Nakanishi, K. Fujita, K. Horita, and H. Takahashi, "Terahertz imaging with room-temperature terahertz difference-frequency quantum-cascade laser sources," *Optics Express*, vol. 27, no. 3, pp. 1884–1893, 2019.

[29] J. Beeman and E. Haller, "Ge:Ga photoconductor arrays: Design considerations and quantitative analysis of prototype single pixels," *Infrared Physics & Technology*, vol. 35, no. 7, pp. 827–836, 1994.

[30] J. Nazdrowicz, "Modelling Microbolometer using Matlab/SIMULINK package with thermal noise sources," in *2016 MIXDES - 23rd International Conference Mixed Design of Integrated Circuits and Systems*, pp. 266–270, 2016.

[31] A. Rogalski and F. Sizov, "Terahertz detectors and focal plane arrays," *Opto-electronics review*, vol. 19, no. 3, pp. 346–404, 2011.

[32] Y.-S. Lee, *Principles of terahertz science and technology*, vol. 170. Springer Science & Business Media, 2009.

[33] J. Birch, J. Dromey, and J. Lesurf, "The optical constants of some common low-loss polymers between 4 and 40 cm^{-1}," *Infrared Physics*, vol. 21, no. 4, pp. 225–228, 1981.

[34] F. Sanjuan and J. O. Tocho, "Optical properties of silicon, sapphire, silica and glass in the Terahertz range," in *Latin America Optics and Photonics Conference*, p. LT4C.1, Optical Society of America, 2012.

[35] Tydex, "THz Waveplates." `http://www.tydexoptics.com/products/thz_optics/thz_waveplate`, Accesed: 09–01–2020.

[36] N. Krumbholz, K. Gerlach, F. Rutz, M. Koch, R. Piesiewicz, T. Kürner, and D. Mittleman, "Omnidirectional terahertz mirrors: A key element for future terahertz communication systems," *Applied Physics Letters*, vol. 88, no. 20, p. 202905, 2006.

[37] M. Tecimer, K. Holldack, and L. R. Elias, "Dynamically tunable mirrors for THz free electron laser applications," *Physical Review Special Topics-Accelerators and Beams*, vol. 13, p. 030703, 2010.

[38] H. Ye, Y. Zhang, and W. Shen, "Carrier transport and optical properties in GaAs far-infrared/terahertz mirror structures," *Thin Solid Films*, vol. 514, no. 1, pp. 310–315, 2006.

[39] A. Sengupta, A. Bandyopadhyay, B. Bowden, J. Harrington, and J. Federici, "Characterisation of olefin copolymers using terahertz spectroscopy," *Electronics Letters*, vol. 42, no. 25, pp. 1477–1479, 2006.

[40] A. Podzorov and G. Gallot, "Low-loss polymers for terahertz applications," *Applied Optics*, vol. 47, no. 18, pp. 3254–3257, 2008.

[41] L. Esaki and R. Tsu, "Superlattice and Negative Differential Conductivity in Semiconductors," *IBM Journal of Research and Development*, vol. 14, no. 1, pp. 61–65, 1970.

[42] R. Kazarinov, "Possibility of amplification of electromagnetic waves in a semiconductor with superlattice," *Soviet Physics–Semiconductors*, vol. 5, no. 4, pp. 707–709, 1971.

[43] A. Cho, "How molecular beam epitaxy (MBE) began and its projection into the future," *Journal of Crystal Growth*, vol. 201-202, pp. 1–7, 1999.

[44] Y. Bai, S. Slivken, S. Kuboya, S. R. Darvish, and M. Razeghi, "Quantum cascade lasers that emit more light than heat," *Nature Photonics*, vol. 4, no. 2, p. 99, 2010.

[45] J. Faist, F. Capasso, D. L. Sivco, C. Sirtori, A. L. Hutchinson, and A. Y. Cho, "Quantum cascade laser," *Science*, vol. 264, no. 5158, pp. 553–556, 1994.

[46] J. Faist, F. Capasso, C. Sirtori, D. L. Sivco, J. N. Baillargeon, A. L. Hutchinson, S.-N. G. Chu, and A. Y. Cho, "High power mid-infrared ($\lambda \sim 5$ μm) quantum cascade lasers operating above room temperature," *Applied Physics Letters*, vol. 68, no. 26, pp. 3680–3682, 1996.

[47] C. Sirtori, J. Faist, F. Capasso, D. Sivco, A. L. Hutchinson, and A. Y. Cho, "Long wavelength infrared ($\lambda \approx 11$ μm) quantum cascade lasers," *Applied Physics Letters*, vol. 69, no. 19, pp. 2810–2812, 1996.

[48] M. Beck, D. Hofstetter, T. Aellen, J. Faist, U. Oesterle, M. Ilegems, E. Gini, and H. Melchior, "Continuous wave operation of a mid-infrared semiconductor laser at room temperature," *Science*, vol. 295, no. 5553, pp. 301–305, 2002.

[49] P. Q. Liu, A. J. Hoffman, M. D. Escarra, K. J. Franz, J. B. Khurgin, Y. Dikmelik, X. Wang, J.-Y. Fan, and C. F. Gmachl, "Highly power-efficient quantum cascade lasers," *Nature Photonics*, vol. 4, no. 2, p. 95, 2010.

[50] D. Botez, J. D. Kirch, C. Boyle, K. M. Oresick, C. Sigler, H. Kim, B. B. Knipfer, J. H. Ryu, D. Lindberg, T. Earles, *et al.*, "High-efficiency, high-power mid-infrared quantum cascade lasers," *Optical Materials Express*, vol. 8, no. 5, pp. 1378–1398, 2018.

[51] R. Köhler, A. Tredicucci, F. Beltram, H. E. Beere, E. H. Linfield, A. G. Davies, D. A. Ritchie, R. C. Iotti, and F. Rossi, "Terahertz semiconductor-heterostructure laser," *Nature*, vol. 417, no. 6885, pp. 156–159, 2002.

[52] L. Bosco, M. Franckié, G. Scalari, M. Beck, A. Wacker, and J. Faist, "Thermoelectrically cooled THz quantum cascade laser operating up to 210 K," *Applied Physics Letters*, vol. 115, no. 1, p. 010601, 2019.

[53] M. Wienold, B. Röben, L. Schrottke, R. Sharma, A. Tahraoui, K. Biermann, and H. Grahn, "High-temperature, continuous-wave operation of terahertz quantum-cascade lasers with metal-metal waveguides and third-order distributed feedback," *Optics Express*, vol. 22, pp. 3334–3348, 2014.

[54] J. Faist, *Quantum cascade lasers*. OUP Oxford, 2013.

[55] F. Valmorra, *Terahertz Spectroscopy of Two-Dimensional Nanostructures*. PhD thesis, ETH Zurich, 2016.

[56] C. Sirtori and R. Teissier, "Quantum cascade lasers: overview of basic principles of operation and state of the art," 2006.

[57] B. S. Williams, "Terahertz quantum-cascade lasers," *Nature Photonics*, vol. 1, no. 9, p. 517, 2007.

[58] M. Wienold, L. Schrottke, M. Giehler, R. Hey, W. Anders, and H. T. Grahn, "Low-voltage terahertz quantum-cascade lasers based on lo-phonon-assisted interminiband transitions," *Electronics Letters*, vol. 45, no. 20, pp. 1030–1031, 2009.

[59] M. A. Belkin and F. Capasso, "New frontiers in quantum cascade lasers: high performance room temperature terahertz sources," *Physica Scripta*, vol. 90, no. 11, p. 118002, 2015.

[60] P. Gellie, W. Maineult, A. Andronico, G. Leo, C. Sirtori, S. Barbieri, Y. Chassagneux, J. R. Coudevylle, R. Colombelli, S. P. Khanna, E. H. Linfield, and A. G. Davies, "Effect of transverse mode structure on the far field pattern of metal-metal terahertz quantum cascade lasers," *Journal of Applied Physics*, vol. 104, no. 12, p. 124513, 2008.

[61] F. Xie, C. Caneau, H. P. LeBlanc, S. Coleman, M.-T. Ho, L. C. Hughes, and C.-e. Zah, "Room temperature continuous wave operation of long wavelength (9–11 μm) distributed feedback quantum cascade lasers for glucose detection,"

in *Novel In-Plane Semiconductor Lasers XII*, vol. 8640, p. 864016, International Society for Optics and Photonics, 2013.

[62] Q. Lu, Y. Bai, N. Bandyopadhyay, S. Slivken, and M. Razeghi, "Room-temperature continuous wave operation of distributed feedback quantum cascade lasers with watt-level power output," *Applied Physics Letters*, vol. 97, no. 23, p. 231119, 2010.

[63] Z. Loghmari, M. Bahriz, A. Meguekam, H. Nguyen Van, R. Teissier, and A. N. Baranov, "Continuous wave operation of InAs-based quantum cascade lasers at 20 μm," *Applied Physics Letters*, vol. 115, no. 15, p. 151101, 2019.

[64] M. Franckié, L. Bosco, M. Beck, C. Bonzon, E. Mavrona, G. Scalari, A. Wacker, and J. Faist, "Two-well quantum cascade laser optimization by non-equilibrium Green's function modelling," *Applied Physics Letters*, vol. 112, no. 2, p. 021104, 2018.

[65] M. S. Vitiello, G. Scalari, B. Williams, and P. De Natale, "Quantum cascade lasers: 20 years of challenges," *Optics Express*, vol. 23, no. 4, pp. 5167–5182, 2015.

[66] M. A. Kainz, M. P. Semtsiv, G. Tsianos, S. Kurlov, W. T. Masselink, S. Schönhuber, H. Detz, W. Schrenk, K. Unterrainer, G. Strasser, *et al.*, "Thermoelectric-cooled terahertz quantum cascade lasers," *Optics Express*, vol. 27, no. 15, pp. 20688–20693, 2019.

[67] S. Kumar, C. W. I. Chan, Q. Hu, and J. L. Reno, "A 1.8-THz quantum cascade laser operating significantly above the temperature of $\hbar\omega$/kB," *Nature Physics*, vol. 7, p. 166, 2010. Article.

[68] C. Walther, M. Fischer, G. Scalari, R. Terazzi, N. Hoyler, and J. Faist, "Quantum cascade lasers operating from 1.2 to 1.6 THz," *Applied Physics Letters*, vol. 91, no. 13, p. 131122, 2007.

[69] W. Demtröder and M. Roach, *Laser Spectroscopy: Basic Concepts and Instrumentation*. Advanced Texts in Physics, Springer Berlin Heidelberg, 2013.

[70] X. Huang and Y. L. Yung, "A common misunderstanding about the Voigt line profile," *Journal of the Atmospheric Sciences*, vol. 61, no. 13, pp. 1630–1632, 2004.

[71] O. DeLange, "Optical heterodyne detection," *IEEE Spectrum*, vol. 5, no. 10, pp. 77–85, 1968.

[72] N. S. Kopeika, *A system engineering approach to imaging*. SPIE Optical Engineering Press Bellingham, 1998.

[73] A. Maestrini, B. Thomas, H. Wang, C. Jung, J. Treuttel, Y. Jin, G. Chattopadhyay, I. Mehdi, and G. Beaudin, "Schottky diode-based terahertz frequency multipliers and mixers," *Comptes Rendus Physique*, vol. 11, no. 7-8, pp. 480–495, 2010.

[74] R. Blundell and D. Winkler, "The Superconductor Insulator Superconductor Mixer Receiver–A review," in *Nonlinear Superconductive Electronics and Josephson Devices*, pp. 55–72, Springer, 1991.

[75] S. Cherednichenko, P. Khosropanah, E. Kollberg, M. Kroug, and H. Merkel, "Terahertz superconducting hot-electron bolometer mixers," *Physica C: Superconductivity*, vol. 372, pp. 407–415, 2002.

[76] G. C. Bjorklund, M. Levenson, W. Lenth, and C. Ortiz, "Frequency modulation (FM) spectroscopy," *Applied Physics B*, vol. 32, no. 3, pp. 145–152, 1983.

[77] R. Eichholz, H. Richter, M. Wienold, L. Schrottke, R. Hey, H. Grahn, and H.-W. Hübers, "Frequency modulation spectroscopy with a THz quantum-cascade laser," *Optics Express*, vol. 21, no. 26, pp. 32199–32206, 2013.

[78] J. Hall, L. Hollberg, T. Baer, and H. Robinson, "Optical heterodyne saturation spectroscopy," *Applied Physics Letters*, vol. 39, no. 9, pp. 680–682, 1981.

[79] G. B. Rieker, J. B. Jeffries, and R. K. Hanson, "Calibration-free wavelength-modulation spectroscopy for measurements of gas temperature and concentration in harsh environments," *Applied Optics*, vol. 48, no. 29, pp. 5546–5560, 2009.

[80] S. Svanberg, "Atomic spectroscopy by resonance scattering," *Philosophical Transactions of the Royal Society of London. Series A, Mathematical and Physical Sciences*, vol. 293, no. 1402, pp. 215–222, 1979.

[81] T. K. Instruments, "THz Absolute Power & Energy Meter System." `http://www.terahertz.co.uk/tk-instruments/products/absolute-thz-power-energy-meters`, Accesed: 09–01–2020.

[82] H. Richter, N. Rothbart, and H.-W. Hübers, "Characterizing the beam properties of terahertz quantum-cascade lasers," *Journal of Infrared, Millimeter, and Terahertz Waves*, vol. 35, no. 8, pp. 686–698, 2014.

[83] G. A. Blake, "Microwave and terahertz spectroscopy," *Encyclopedia of Chemical Physics and Physical Chemistry*, vol. 2, pp. 1063–1088, 2001.

[84] H.-W. Hübers, S. Pavlov, H. Richter, A. Semenov, L. Mahler, A. Tredicucci, H. Beere, and D. Ritchie, "High-resolution gas phase spectroscopy with a distributed feedback terahertz quantum cascade laser," *Applied Physics Letters*, vol. 89, no. 6, p. 061115, 2006.

[85] E. R. Hudson, H. Lewandowski, B. C. Sawyer, and J. Ye, "Cold molecule spectroscopy for constraining the evolution of the fine structure constant," *Physical Review Letters*, vol. 96, no. 14, p. 143004, 2006.

[86] C. Risacher, R. Güsten, J. Stutzki, H.-W. Hübers, R. Aladro, A. Bell, C. Buchbender, D. Büchel, T. Csengeri, C. Duran, U. U. Graf, R. D. Higgins, C. E. Honingh, K. Jacobs, M. Justen, B. Klein, M. Mertens, Y. Okada, A. Parikka, P. Pütz, N. Reyes, H. Richter, O. Ricken, D. Riquelme, N. Rothbart, N. Schneider, R. Simon, M. Wienold, H. Wiesemeyer, M. Ziebart, P. Fusco, S. Rosner, and B. Wohler, "The upGREAT Dual Frequency Heterodyne Arrays for SOFIA," *J. Astron. Instrum.*, vol. 7, no. 04, p. 1840014, 2018.

[87] L. Rezac, P. Hartogh, R. Güsten, H. Wiesemeyer, H.-W. Hübers, C. Jarchow, H. Richter, B. Klein, and N. Honingh, "First detection of the 63 μm atomic oxygen line in the thermosphere of Mars with GREAT/SOFIA," *Astronomy & Astrophysics*, vol. 580, p. L10, 2015.

[88] X. Wang, C. Shen, T. Jiang, Z. Zhan, Q. Deng, W. Li, W. Wu, N. Yang, W. Chu, and S. Duan, "High-power terahertz quantum cascade lasers with $\sim$0.23 W in continuous wave mode," *AIP Advances*, vol. 6, no. 7, p. 075210, 2016.

[89] M. S. Vitiello, L. Consolino, S. Bartalini, A. Taschin, A. Tredicucci, M. Inguscio, and P. De Natale, "Quantum-limited frequency fluctuations in a terahertz laser," *Nature Photonics*, vol. 6, no. 8, p. 525, 2012.

[90] H. M. Pickett, R. L. Poynter, E. A. Cohen, M. L. Delitsky, J. C. Pearson, and H. S. P. Müller, "Submillimeter, millimeter, and microwave spectral line catalog," *Journal of Quantitative Spectroscopy and Radiative Transfer*, vol. 60, no. 5, pp. 883–890, 1998.

[91] J. T. Remillard, D. Uy, W. H. Weber, F. Capasso, C. Gmachl, A. Hutchinson, D. Sivco, J. Baillargeon, and A. Cho, "Sub-Doppler resolution limited Lamb-dip

spectroscopy of NO with a quantum cascade distributed feedback laser," *Optics Express*, vol. 7, no. 7, pp. 243–248, 2000.

[92] N. Mukherjee, R. Go, and C. K. N. Patel, "Linewidth measurement of external grating cavity quantum cascade laser using saturation spectroscopy," *Applied Physics Letters*, vol. 92, no. 11, p. 111116, 2008.

[93] R. Walker, J. Kirkbride, J. van Helden, D. Weidmann, and G. Ritchie, "Sub-Doppler spectroscopy with an external cavity quantum cascade laser," *Applied Physics B*, vol. 112, no. 2, pp. 159–167, 2013.

[94] J. C. Pearson, B. J. Drouin, A. Maestrini, I. Mehdi, J. Ward, R. H. Lin, S. Yu, J. J. Gill, B. Thomas, C. Lee, *et al.*, "Demonstration of a room temperature 2.48–2.75 THz coherent spectroscopy source," *Review of Scientific Instruments*, vol. 82, no. 9, p. 093105, 2011.

[95] G. Cazzoli and C. Puzzarini, "Sub-Doppler resolution in the THz frequency domain: 1 kHz accuracy at 1 THz by exploiting the Lamb-dip technique," *The Journal of Physical Chemistry A*, vol. 117, no. 50, pp. 13759–13766, 2013.

[96] L. Consolino, A. Campa, M. Ravaro, D. Mazzotti, M. Vitiello, S. Bartalini, and P. De Natale, "Saturated absorption in a rotational molecular transition at 2.5 THz using a quantum cascade laser," *Applied Physics Letters*, vol. 106, no. 2, p. 021108, 2015.

[97] V. S. Letokhov and V. P. Chebotayev, *Nonlinear laser spectroscopy*, vol. 4. Springer, 1977.

[98] M. Wienold, T. Alam, L. Schrottke, H. Grahn, and H.-W. Hübers, "Doppler-free spectroscopy with a terahertz quantum-cascade laser," *Optics Express*, vol. 26, no. 6, pp. 6692–6699, 2018.

[99] Y. J. Han, J. Partington, R. Chhantyal-Pun, M. Henry, O. Auriacombe, T. Rawlings, L. H. Li, J. Keeley, M. Oldfield, N. Brewster, R. Dong, P. Dean, A. G. Davies, B. N. Ellison, E. H. Linfield, and A. Valavanis, "Gas spectroscopy through multimode self-mixing in a double-metal terahertz quantum cascade laser," *Optics Letters*, vol. 43, no. 24, pp. 5933–5936, 2018.

[100] T. Hagelschuer, M. Wienold, H. Richter, L. Schrottke, H. T. Grahn, and H.-W. Hübers, "Real-time gas sensing based on optical feedback in a terahertz quantum-cascade laser," *Optics Express*, vol. 25, no. 24, pp. 30203–30213, 2017.

[101] J. Xu, J. M. Hensley, D. Fenner, R. P. Green, L. Mahler, A. Tredicucci, M. G. Allen, F. Beltram, H. E. Beere, and D. A. Ritchie, "Tunable terahertz quantum cascade lasers with an external cavity," *Applied Physics Letters*, vol. 91, no. 12, p. 121104, 2007.

[102] A. W. M. Lee, B. S. Williams, S. Kumar, Q. Hu, and J. L. Reno, "Tunable terahertz quantum cascade lasers with external gratings," *Optics Letters*, vol. 35, no. 7, pp. 910–912, 2010.

[103] Q. Qin, B. S. Williams, S. Kumar, J. L. Reno, and Q. Hu, "Tuning a terahertz wire laser," *Nature Photonics*, vol. 3, no. 12, p. 732, 2009.

[104] N. Han, A. de Geofroy, D. P. Burghoff, C. W. I. Chan, A. W. M. Lee, J. L. Reno, and Q. Hu, "Broadband all-electronically tunable MEMS terahertz quantum cascade lasers," *Optics Letters*, vol. 39, no. 12, pp. 3480–3483, 2014.

[105] K. Ohtani, M. Beck, and J. Faist, "Electrical laser frequency tuning by three terminal terahertz quantum cascade lasers," *Applied Physics Letters*, vol. 104, no. 1, p. 011107, 2014.

[106] I. Kundu, P. Dean, A. Valavanis, L. Chen, L. Li, J. E. Cunningham, E. H. Linfield, and A. G. Davies, "Discrete Vernier tuning in terahertz quantum cascade lasers using coupled cavities," *Optics Express*, vol. 22, no. 13, pp. 16595–16605, 2014.

[107] I. Kundu, P. Dean, A. Valavanis, J. R. Freeman, M. C. Rosamond, L. Li, Y. Han, E. H. Linfield, and A. G. Davies, "Continuous frequency tuning with near constant output power in coupled y-branched terahertz quantum cascade lasers with photonic lattice," *ACS Photonics*, vol. 5, no. 7, pp. 2912–2920, 2018.

[108] D. Turčinková, M. Ines Amanti, F. Castellano, M. Beck, and J. Faist, "Continuous tuning of terahertz distributed feedback quantum cascade laser by gas condensation and dielectric deposition," *Applied Physics Letters*, vol. 102, no. 18, p. 181113, 2013.

[109] L. A. Dunbar, R. Houdré, G. Scalari, L. Sirigu, M. Giovannini, and J. Faist, "Small optical volume terahertz emitting microdisk quantum cascade lasers," *Applied Physics Letters*, vol. 90, no. 14, p. 141114, 2007.

[110] G. Chen, R. Martini, S.-w. Park, C. G. Bethea, I.-C. A. Chen, P. D. Grant, R. Dudek, and H. C. Liu, "Optically induced fast wavelength modulation in a quantum cascade laser," *Applied Physics Letters*, vol. 97, no. 1, p. 011102, 2010.

[111] S. Suchalkin, S. Jung, R. Tober, M. Belkin, and G. Belenky, "Optically tunable long wavelength infrared quantum cascade laser operated at room temperature," *Applied Physics Letters*, vol. 102, no. 1, p. 011125, 2013.

[112] D. Guo, H. Cai, M. A. Talukder, X. Chen, A. M. Johnson, J. B. Khurgin, and F.-S. Choa, "Near-infrared induced optical quenching effects on mid-infrared quantum cascade lasers," *Applied Physics Letters*, vol. 104, no. 25, p. 251102, 2014.

[113] M. Hempel, B. Röben, L. Schrottke, H.-W. Hübers, and H. T. Grahn, "Fast continuous tuning of terahertz quantum-cascade lasers by rear-facet illumination," *Applied Physics Letters*, vol. 108, no. 19, p. 191106, 2016.

[114] S. L. Chuang, *Physics of photonic devices*, vol. 80. John Wiley & Sons, 2012.

[115] C. Henry and D. V. Lang, "Nonradiative capture and recombination by multiphonon emission in GaAs and GaP," *Physical Review B*, vol. 15, no. 2, p. 989, 1977.

[116] Y. Varshni, "Band-to-band radiative recombination in groups IV, VI, and III-V semiconductors (I)," *Physica Status Solidi (B)*, vol. 19, no. 2, pp. 459–514, 1967.

[117] M. Born and E. Wolf, "Principles of Optics, 7th (expanded) edition," *United Kingdom: Press Syndicate of the University of Cambridge*, vol. 461, 1999.

[118] M. Levinshtein, S. Rumyantsev, and M. Shur, *Handbook Series on Semiconductor Parameters*. World Scientific, 1996.

[119] M. Rösch, G. Scalari, M. Beck, and J. Faist, "Octave-spanning semiconductor laser," *Nature Photonics*, vol. 9, no. 1, p. 42, 2015.

[120] M. Ravaro, S. Barbieri, G. Santarelli, V. Jagtap, C. Manquest, C. Sirtori, S. P. Khanna, and E. H. Linfield, "Measurement of the intrinsic linewidth of terahertz quantum cascade lasers using a near-infrared frequency comb," *Optics Express*, vol. 20, no. 23, pp. 25654–25661, 2012.

[121] H.-W. Hübers, H. Richter, R. Eichholz, M. Wienold, K. Biermann, L. Schrottke, and H. T. Grahn, "Heterodyne Spectroscopy of Frequency Instabilities in Terahertz Quantum-Cascade Lasers Induced by Optical Feedback," *IEEE Journal of Selected Topics in Quantum Electronics*, vol. 23, no. 4, p. 1800306, 2017.

[122] A. Barkan, F. K. Tittel, D. M. Mittleman, R. Dengler, P. H. Siegel, G. Scalari, L. Ajili, J. Faist, H. E. Beere, E. H. Linfield, A. G. Davies, and D. A. Ritchie,

"Linewidth and tuning characteristics of terahertz quantum cascade lasers," *Optics Letters*, vol. 29, no. 6, pp. 575–577, 2004.

[123] S. Barbieri, J. Alton, H. E. Beere, E. H. Linfield, D. A. Ritchie, S. Withington, G. Scalari, L. Ajili, and J. Faist, "Heterodyne mixing of two far-infrared quantum cascade lasers by use of a point-contact Schottky diode," *Optics Letters*, vol. 29, no. 14, pp. 1632–1634, 2004.

[124] P. Khosropanah, A. Baryshev, W. Zhang, W. Jellema, J. N. Hovenier, J. R. Gao, T. M. Klapwijk, D. G. Paveliev, B. S. Williams, S. Kumar, Q. Hu, J. L. Reno, B. Klein, and J. L. Hesler, "Phase locking of a 2.7 THz quantum cascade laser to a microwave reference," *Optics Letters*, vol. 34, no. 19, pp. 2958–2960, 2009.

[125] D. Rabanus, U. U. Graf, M. Philipp, O. Ricken, J. Stutzki, B. Vowinkel, M. C. Wiedner, C. Walther, M. Fischer, and J. Faist, "Phase locking of a 1.5 Terahertz quantum cascade laser and use as a local oscillator in a heterodyne HEB receiver," *Optics Express*, vol. 17, no. 3, pp. 1159–1168, 2009.

[126] A. Danylov, N. Erickson, A. Light, and J. Waldman, "Phase locking of 2.324 and 2.959 terahertz quantum cascade lasers using a Schottky diode harmonic mixer," *Optics Letters*, vol. 40, no. 21, pp. 5090–5092, 2015.

[127] L. Consolino, A. Taschin, P. Bartolini, S. Bartalini, P. Cancio, A. Tredicucci, H. E. Beere, D. A. Ritchie, R. Torre, M. S. Vitiello, and P. De Natale, "Phase-locking to a free-space terahertz comb for metrological-grade terahertz lasers," *Nature Communications*, vol. 3, no. 1, p. 1040, 2012.

[128] M. Ravaro, C. Manquest, C. Sirtori, S. Barbieri, G. Santarelli, K. Blary, J.-F. Lampin, S. Khanna, and E. Linfield, "Phase-locking of a 2.5 THz quantum cascade laser to a frequency comb using a GaAs photomixer," *Optics Letters*, vol. 36, no. 20, pp. 3969–3971, 2011.

[129] A. L. Betz, R. T. Boreiko, B. S. Williams, S. Kumar, Q. Hu, and J. L. Reno, "Frequency and phase-lock control of a 3 THz quantum cascade laser," *Optics Letters*, vol. 30, no. 14, pp. 1837–1839, 2005.

[130] A. A. Danylov, T. M. Goyette, J. Waldman, M. J. Coulombe, A. J. Gatesman, R. H. Giles, W. D. Goodhue, X. Qian, and W. E. Nixon, "Frequency stabilization of a single mode terahertz quantum cascade laser to the kilohertz level," *Optics Express*, vol. 17, no. 9, pp. 7525–7532, 2009.

[131] J. R. Freeman, L. Ponnampalam, H. Shams, R. A. Mohandas, C. C. Renaud, P. Dean, L. Li, A. G. Davies, A. J. Seeds, and E. H. Linfield, "Injection locking of a terahertz quantum cascade laser to a telecommunications wavelength frequency comb," *Optica*, vol. 4, no. 9, pp. 1059–1064, 2017.

[132] A. Arie, S. Schiller, E. K. Gustafson, and R. L. Byer, "Absolute frequency stabilization of diode-laser-pumped Nd:YAG lasers to hyperfine transitions in molecular iodine," *Optics Letters*, vol. 17, no. 17, pp. 1204–1206, 1992.

[133] G. Galzerano, C. Svelto, E. Bava, and F. Bertinetto, "High-frequency-stability diode-pumped Nd:YAG lasers with the FM sidebands method and Doppler-free iodine lines at 532 nm," *Applied Optics*, vol. 38, no. 33, pp. 6962–6966, 1999.

[134] H. Richter, S. G. Pavlov, A. D. Semenov, L. Mahler, A. Tredicucci, H. E. Beere, D. A. Ritchie, and H.-W. Hübers, "Submegahertz frequency stabilization of a terahertz quantum cascade laser to a molecular absorption line," *Applied Physics Letters*, vol. 96, no. 7, p. 071112, 2010.

[135] F. Cappelli, I. Galli, S. Borri, G. Giusfredi, P. Cancio, D. Mazzotti, A. Montori, N. Akikusa, M. Yamanishi, S. Bartalini, and P. D. Natale, "Subkilohertz linewidth room-temperature mid-infrared quantum cascade laser using a molecular sub-Doppler reference," *Optics Letters*, vol. 37, no. 23, pp. 4811–4813, 2012.

[136] Y. Ren, D. J. Hayton, J. N. Hovenier, M. Cui, J. R. Gao, T. M. Klapwijk, S. C. Shi, T.-Y. Kao, Q. Hu, and J. L. Reno, "Frequency and amplitude stabilized terahertz quantum cascade laser as local oscillator," *Applied Physics Letters*, vol. 101, no. 10, p. 101111, 2012.

[137] B. Wei, S. J. Kindness, N. W. Almond, R. Wallis, Y. Wu, Y. Ren, S. C. Shi, P. Braeuninger-Weimer, S. Hofmann, H. E. Beere, D. A. Ritchie, and R. Degl'Innocenti, "Amplitude stabilization and active control of a terahertz quantum cascade laser with a graphene loaded split-ring-resonator array," *Applied Physics Letters*, vol. 112, no. 20, p. 201102, 2018.

List of publications

Publications

1. T. Alam, M. Wienold, X. Lü, K. Biermann, L. Schrottke, H. T. Grahn, and H.-W. Hübers, *Frequency and power stabilization of a terahertz quantum-cascade laser using near-infrared optical excitation*, Opt. Express 27, 36846–36854 (2019).

2. T. Alam, M. Wienold, X. Lü, K. Biermann, L. Schrottke, H. T. Grahn, and H.-W. Hübers, *Wideband, high-resolution terahertz spectroscopy by light-induced frequency tuning of quantum-cascade lasers*, Opt. Express 27, 5420–5432 (2019).

3. M. Wienold, T. Alam, L. Schrottke, H. T. Grahn, and H.-W. Hübers, *Doppler-free spectroscopy with a terahertz quantum-cascade laser*, Opt. Express 26, 6692–6699 (2018).

Conference presentations

1. T. Alam, M. Wienold, X. Lü, K. Biermann, L. Schrottke, H. T. Grahn, and H.-W. Hübers, *Stabilizing a terahertz quantum-cascade laser using near-infrared optical excitation*, "44th International Conference on Infrared, Millimeter, and Terahertz Waves (IRMMW-THz)", Paris, France, September, 2019.

2. T. Alam, M. Wienold, X. Lü, K. Biermann, L. Schrottke, H. T. Grahn, and H.-W. Hübers, *Frequency tuning of terahertz quantum-cascade lasers by spatially controlled optical excitation*, "French-German THz Conference (FGTC-2019)", Kaiserslautern, Germany, April, 2019.

3. T. Alam, M. Wienold, L. Schrottke, H. T. Grahn, and H.-W. Hübers, *Molecular spectroscopy with a terahertz quantum-cascade laser by illumination-induced frequency tuning*, "8th International Quantum Cascade Laser School and Workshop (IQCLSW)" Cassis, France, September, 2018.

4. M. Wienold, T. Alam, L. Schrottke, H. T. Grahn, and H.-W. Hübers, *Lamb dip spectroscopy with a terahertz quantum-cascade laser*, "42nd International Conference on Infrared, Millimeter, and Terahertz Waves (IRMMW-THz)", Cancun, Mexico, August, 2017.

Acknowledgements

None of this fascinating research would have been possible without the support of a large group of people. I would like to take a moment to express my gratitude to thank those people who have made this thesis possible.

I am very grateful to my two Ph.D. mentors, Prof. Dr. Heinz-Wilhelm Hübers and Dr. Martin Weinold, who together have guided my last three and a half years of efforts to this end. In Particular, I wish to express my deep and sincere gratitude to my research advisor, Prof. Dr. Heinz-Wilhelm Hübers. I've been inspired by his high working morale, his passion, and his unique leadership. I wish that along my personal and professional journey I could develop some of those characteristics. Second but not least, my main contributor is my dear supervisor, Dr. Martin Wienold, for whom I am indebted and forever grateful. He has taught me how to deal with research laboratories and experimental analysis. He was always available to figure out and discuss any doubts that might have emerged in my mind.

Of course, I must thank the person who made this possible, Prof. Günther Tränkle, my main supervisor. When I was struggling to find my supervisor, he took me as his Ph.D. student and reviewed my work regularly.

I would like to thank my co-authors, Prof. Holger. T. Grahn and Dr. Lutz Schrottke, and Dr. Xiang Lü for reading my conference and research papers carefully and for teaching me how to write good scientific papers.

I would like to thank all colleagues in the THz laser spectroscopy group for the excellent working atmosphere. The whole group atmosphere was very optimistic, supportive, and inspiring, which helped a lot. I'm not going to try mentioning names for fear of leaving someone out, but I can assure you that you've all been very helpful on this journey.

My dear family has been a great support to me and has always been at my side with their affection, without making me feel the stress of living abroad. Last but not least, the biggest thanks to my life companion, Hima: after all these years of long-distance relationships, we are together!! I will be indebted to you for the level of support you gave me during my bad days. Without all your help, and your beautiful smile, I couldn't have managed it.

Innovationen mit Mikrowellen und Licht
Forschungsberichte aus dem Ferdinand-Braun-Institut, Leibniz-Institut für Höchstfrequenztechnik

Herausgeber: Prof. Dr. G. Tränkle, Prof. Dr.-Ing. W. Heinrich

Band 1: **Thorsten Tischler**
Die Perfectly-Matched-Layer-Randbedingung in der Finite-Differenzen-Methode im Frequenzbereich: Implementierung und Einsatzbereiche
ISBN: 3-86537-113-2, 19,00 EUR, 144 Seiten

Band 2: **Friedrich Lenk**
Monolithische GaAs FET- und HBT-Oszillatoren mit verbesserter Transistormodellierung
ISBN: 3-86537-107-8, 19,00 EUR, 140 Seiten

Band 3: **R. Doerner, M. Rudolph (eds.)**
Selected Topics on Microwave Measurements, Noise in Devices and Circuits, and Transistor Modeling
ISBN: 3-86537-328-3, 19,00 EUR, 130 Seiten

Band 4: **Matthias Schott**
Methoden zur Phasenrauschverbesserung von monolithischen Millimeterwellen-Oszillatoren
ISBN: 978-3-86727-774-0, 19,00 EUR, 134 Seiten

Band 5: **Katrin Paschke**
Hochleistungsdiodenlaser hoher spektraler Strahldichte mit geneigtem Bragg-Gitter als Modenfilter (α-DFB-Laser)
ISBN: 978-3-86727-775-7, 19,00 EUR, 128 Seiten

Band 6: **Andre Maaßdorf**
Entwicklung von GaAs-basierten Heterostruktur-Bipolartransistoren (HBTs) für Mikrowellenleistungszellen
ISBN: 978-3-86727-743-3, 23,00 EUR, 154 Seiten

Band 7: **Prodyut Kumar Talukder**
Finite-Difference-Frequency-Domain Simulation of Electrically Large Microwave Structures using PML and Internal Ports
ISBN: 978-3-86955-067-1, 19,00 EUR, 138 Seiten

Band 8: **Ibrahim Khalil**
Intermodulation Distortion in GaN HEMT
ISBN: 978-3-86955-188-3, 23,00 EUR, 158 Seiten

Band 9: **Martin Maiwald**
Halbleiterlaser basierte Mikrosystemlichtquellen für die Raman-Spektroskopie
ISBN: 978-3-86955-184-5, 19,00 EUR, 134 Seiten

Band 10: **Jens Flucke**
Mikrowellen-Schaltverstärker in GaN- und GaAs-Technologie Designgrundlagen und Komponenten
ISBN: 978-3-86955-304-7, 21,00 EUR, 122 Seiten

Cuvillier Verlag
Internationaler wissenschaftlicher Fachverlag

Innovationen mit Mikrowellen und Licht
Forschungsberichte aus dem Ferdinand-Braun-Institut, Leibniz-Institut für Höchstfrequenztechnik

Herausgeber: Prof. Dr. G. Tränkle, Prof. Dr.-Ing. W. Heinrich

Band 11: **Harald Klockenhoff**
Optimiertes Design von Mikrowellen-Leistungstransistoren und Verstärkem im X-Band
ISBN: 978-3-86955-391-7, 26,75 EUR, 130 Seiten

Band 12: **Reza Pazirandeh**
Monolithische GaAs FET- und HBT-Oszillatoren mit verbesserter Transistormodellierung
ISBN: 978-3-86955-107-8, 19,00 EUR, 140 Seiten

Band 13: **Tomas Krämer**
High-Speed InP Heterojunction Bipolar Transistors and Integrated Circuits in Transferred Substrate Technology
ISBN: 978-3-86955-393-1, 21,70 EUR, 140 Seiten

Band 14: **Phuong Thanh Nguyen**
Investigation of spectral characteristics of solitary diode lasers with integrated grating resonator
ISBN: 978-3-86955-651-2, 24,00 EUR, 156 Seiten

Band 15: **Sina Riecke**
Flexible Generation of Picosecond Laser Pulses in the Infrared and Green Spectral Range by Gain-Switching of Semiconductor Lasers
ISBN: 978-3-86955-652-9, 22,60 EUR, 136 Seiten

Band 16: **Christian Hennig**
Hydrid-Gasphasenepitaxie von versetzungsarmen und freistehenden GaN-Schichten
ISBN: 978-3-86955-822-6, 27,00 EUR, 162 Seiten

Band 17: **Tim Wernicke**
Wachstum von nicht- und semipolaren InAlGaN-Heterostrukturen für hocheffiziente Licht-Emitter
ISBN: 978-3-86955-881-3, 23,40 EUR, 138 Seiten

Band 18: **Andreas Wentzel**
Klasse-S Mikrowellen-Leistungsverstärker mit GaN-Transistoren
ISBN: 978-3-86955-897-4, 29,65 EUR, 172 Seiten

Band 19: **Veit Hoffmann**
MOVPE growth and characterization of (In,Ga)N quantum structures for laser diodes emitting at 440 nm
ISBN: 978-3-86955-989-6, 18,00 EUR, 118 Seiten

Band 20: **Ahmad Ibrahim Bawamia**
Improvement of the beam quality of high-power broad area semiconductor diode lasers by means of an external resonator
ISBN: 978-3-95404-065-0, 21,00 EUR, 126 Seiten

Cuvillier Verlag
Internationaler wissenschaftlicher Fachverlag

Innovationen mit Mikrowellen und Licht
Forschungsberichte aus dem Ferdinand-Braun-Institut, Leibniz-Institut für Höchstfrequenztechnik

Herausgeber: Prof. Dr. G. Tränkle, Prof. Dr.-Ing. W. Heinrich

Band 21: **Agnietzka Pietrzak**
Realization of High Power Diode Lasers with Extremely Narrow Vertical Divergence
ISBN: 978-3-95404-066-7, 27,40 EUR, 144 Seiten

Band 22: **Eldad Bahat-Treidel**
GaN-based HEMTs for High Voltage Operation
Design, Technology and Characterization
ISBN: 978-3-95404-094-0, 41,10 EUR, 220 Seiten

Band 23: **Ponky Ivo**
AlGaN/GaN HEMTs Reliability:
Degradation Modes and Anslysis
ISBN: 978-3-95404-259-3, 23,55 EUR, 132 Seiten

Band 24: **Stefan Spießberger**
Compact Semiconductor-Based Laser Sources
with Narrow Linewidth and High Output Power
ISBN: 978-3-95404-261-6, 24,15 EUR, 140 Seiten

Band 25: **Silvio Kühn**
Mikrowellenoszillatoren für die Erzeugung von atmosphärischen Mikroplasmen
ISBN: 978-3-95404-378-1, 21,85 EUR, 112 Seiten

Band 26: **Sven Schwertfeger**
Experimentelle Untersuchung der Modensynchronisation in Multisegment-Laserdioden zur Erzeugung kurzer optischer Pulse bei einer Wellenlänge von 920 nm
ISBN: 978-3-95404-471-9, 29,45 EUR, 150 Seiten

Band 27: **Christoph Matthias Schultz**
Analysis and mitigation of the factors limiting the effiency of high power distributed feedback diode lasers
ISBN: 978-3-95404-521-1, 68,40 EUR, 388 Seiten

Band 28: **Luca Redaelli**
Design and fabrication of GaN-based laser diodes for single-mode and narrow-linewidth applications
ISBN: 978-3-95404-586-0, 29,70 EUR, 176 Seiten

Band 29: **Martin Spreemann**
Resonatorkonzepte für Hochleistungs-Diodenlaser
mit ausgedehnten lateralen Dimensionen
ISBN: 978-3-95404-628-7, 25,15 EUR, 128 Seiten

Cuvillier Verlag
Internationaler wissenschaftlicher Fachverlag

Innovationen mit Mikrowellen und Licht
Forschungsberichte aus dem Ferdinand-Braun-Institut, Leibniz-Institut für Höchstfrequenztechnik

Herausgeber: Prof. Dr. G. Tränkle, Prof. Dr.-Ing. W. Heinrich

Band 30:	**Christian Fiebig** Diodenlaser mit Trapezstruktur und hoher Brillanz für die Realisierung einer Frequenzkonversion auf einer mikro-optischen Bank ISBN: 978-3-95404-690-4, 26,30 EUR, 140 Seiten
Band 31:	**Viola Küller** Versetzungsreduzierte AlN- und AlGaN-Schichten als Basis für UV LEDs ISBN: 978-3-95404-741-3, 34,40 EUR, 164 Seiten
Band 32:	**Daniel Jedrzejczyk** Efficient frequency doubling of near-infrared diode lasers using quasi phase-matched waveguides ISBN: 978-3-95404-958-5, 27,90 EUR, 134 Seiten
Band 33:	**Sylvia Hagedorn** Hybrid-Gasphasenepitaxie zur Herstellung von Aluminiumgalliumnitrid ISBN: 978-3-95404-985-1, 38,00 EUR, 176 Seiten
Band 34:	**Alexander Kravets** Advanced Silicon MMICs for mm-Wave Automotive Radar Front-Ends ISBN: 978-3-95404-986-8, 31,90 EUR, 156 Seiten
Band 35:	**David Feise** Longitudinale Modenfilter für Kantanemitter im roten Spektralbereich ISBN: 978-3-7369-9116-3, 39,20 EUR, 168 Seiten
Band 36:	**Ksenia Nosaeva** Indium phosphide HBT in thermally optimized periphery for applications up to 300GHZ ISBN: 978-3-7369-287-0, 42,00 EUR, 154 Seiten
Band 37:	**Muhammad Maruf Hossain** Signal Generation for Millimeter Wave and THZ Applications in InP-DHBT and InP-on-BiCMOS Technologies ISBN: 978-3-7369-9335-8, 35,60 EUR, 136 Seiten
Band 38:	**Sirinpa Monayakul** Development of Sub-mm Wave Flip-Chip Interconnect ISBN: 978-3-7369-9410-2, 44,00 EUR, 146 Seiten
Band 39:	**Moritz Brendel** Charakterisierung und Optimierung von (Al, Ga) N-basierten UV-Photodetektoren ISBN: 978-3-7369-9465-2, 49,90 EUR, 196 Seiten
Band 40:	**Erdenetsetseg Luvsandamdin** Development of micro-integrated diode lasers for precision quantum optics experiments in space ISBN: 978-3-7369-9479-9, 39,00 EUR, 126 Seiten

Cuvillier Verlag
Internationaler wissenschaftlicher Fachverlag

Innovationen mit Mikrowellen und Licht

Forschungsberichte aus dem Ferdinand-Braun-Institut, Leibniz-Institut für Höchstfrequenztechnik

Herausgeber: Prof. Dr. G. Tränkle, Prof. Dr.-Ing. W. Heinrich

Band 41: **Thi Nghiem Vu**
Development and analysis of diode laser ns-MOPA systems for high peak power application
ISBN: 978-3-7369-9480-5, 38,80 EUR, 138 Seiten

Band 42: **Christian Bansleben**
Differentieller Mikrowellen-Leistungsoszillator für die Realisierung ultrakompakter Plasmaquellen in Matrixanordnung
ISBN: 978-3-7369-9530-7, 34,90 EUR, 136 Seiten

Band 43: **Martin Winterfeldt**
Investigation of slow-axis beam quality degradation in high-power broad area diode lasers
ISBN: 978-3-7369-9733-2, 39,90 EUR, 158 Seiten

Band 44: **Jonathan Decker**
Investigation of monolithically integrated spectral stabilization in high-brightness broad area diode lasers
ISBN: 978-3-7369-9798-1, 49,50 EUR, 174 Seiten

Band 45: **Andreea Cristina Andrei**
Untersuchung und Optimierung robuster und hochlinearer rauscharmer Verstärker in GaN-Technologie
ISBN: 978-3-7369-9810-0, 39,90 EUR, 148 Seiten

Band 46: **Peng Luo**
GaN HEMT Modeling Including Trapping Effects Based on Chalmers Model and Pulsed S-Parameter Measurements
ISBN: 978-3-7369-9906-0, 48,00 EUR, 160 Seiten

Band 47: **Jörg Jeschke**
Entwicklung von optisch pumpbaren UVC-Lasern auf AlGaN-Basis
Chalmers Model and Pulsed S-Parameter Measurements
ISBN: 978-3-7369-9918-3, 44,90 EUR, 176 Seiten

Band 48: **Nikolai Wolff**
Wideband GaN Microwave Power Amplifiers with Class-G Supply Modulation
ISBN: 978-3-7369-9931-2, 44,90 EUR, 170 Seiten

Band 49: **Carlo Frevert**
Optimization of broad-area GaAs diode lasers for high powers and high efficiences in the temperature range 200-220 K
ISBN: 978-3-7369-9944-2, 44,90 EUR, 174 Seiten

Band 50: **Bassem Arar**
GaAs-based components for photonic integrated circuits
ISBN: 978-3-7369-9976-3, 43,60 EUR, 152 Seiten

Cuvillier Verlag
Internationaler wissenschaftlicher Fachverlag

Innovationen mit Mikrowellen und Licht
Forschungsberichte aus dem Ferdinand-Braun-Institut, Leibniz-Institut für Höchstfrequenztechnik

Herausgeber: Prof. Dr. G. Tränkle, Prof. Dr.-Ing. W. Heinrich

Band 51: **Mahmoud Tawfieq**
Development and characterisation of a diode laser based tunable high-power MOPA system
ISBN: 978-3-7369-9983-1, 44,90 EUR, 170 Seiten

Band 52: **Sebastian Preis**
Hocheffiziente frequenzagile Mikrowellen-Leistungsverstärker auf Basis von Verbindungshalbleitern und Ferroelektrika
ISBN: 978-3-7369-7004-5, 54,00 EUR, 136 Seiten

Band 53: **Simon Fleischmann**
Materialaspekte der Hydridgasphasenepitaxie von Aluminiumgalliumnitrid
ISBN: 978-3-7369-7029-8, 41,80 EUR, 160 Seiten

Band 54: **Erhan Ersoy**
Optimierung von koplanaren GaN-MMIC-Leistungsverstärkern im X-Band
ISBN: 978-3-7369-7157-8, 39,90 EUR, 154 Seiten

Band 55: **Simon Rauch**
Lateral emission characteristics of high-power broad-area lasers subject to external optical feedback
ISBN: 978-3-7369-7168-5, 29,90 EUR, 112 Seiten

Band 56: **Frank Dittmar**
Untersuchung der Strahlgüte von brillanten Hochleistungs-Trapezlasern für den Wellenlängenbereich bei 808 nm
ISBN: 978-3-7369-7167-7, 44,90 EUR, 176 Seiten

Band 57: **Florian Hühn**
Flexibler Modulator und Digitalverstärker-MMIC für den energieeffizienten Betrieb einer digitalen Sendekette im GHz-Bereich
ISBN: 978-3-7369-7190-5, 29,90 EUR, 162 Seiten

Band 58: **Dimitri Stoppel**
Interconnection development for InP-HBT terahertz circuits
ISBN: 978-3-7369-7204-9, 44,90 EUR, 156 Seiten

Band 59: **Anissa Zeghuzi**
Analysis of Spatio-Temporal Phenomena in High-Brightness Diode Lasers using Numerical Simulations
ISBN: 978-3-7369-7289-6, 49,90 EUR, 176 Seiten

Cuvillier Verlag
Internationaler wissenschaftlicher Fachverlag

www.ingramcontent.com/pod-product-compliance
Ingram Content Group UK Ltd.
Pitfield, Milton Keynes, MK11 3LW, UK
UKHW022000190726
13853UKWH00004B/1643

9 783736 972971